THE METRIC DEBATE

The METRIC DEBATE

edited by
David F. Bartlett

Colorado Associated University Press

389.16
m 594

The publishers gratefully acknowledge the
permission of the United States Department
of Commerce, National Bureau of Standards,
to reprint part of NBS Special Publication
330, "The International System of Units," as
the appendix to this book.

Contents

64965

Introduction

The citizens of the United States are deeply divided over the merits of metrication. Some feel that the intrinsic merits of the metric system and its preeminence in world commerce make it advisable to convert our measures as soon as possible. Others believe that conversion will facilitate the setting of improved standards for industry.

Against these hopes are the fears of what the loss of our present system will mean to our language and our perception of quantities. Finally, some feel that the metric system is itself becoming obsolete; that its universal adoption may preclude the development of superior systems of measurement.

These concerns are the subject of this book, which has evolved from a series of talks given at the annual meeting of the American Association for the Advancement of Science in Houston in January, 1979. Presenting the case for metrication are Louis Sokol, the president of the U.S. Metric Association, and Stanley Mallen, the Metrication Planning Manager for the Ford Motor Company. In opposition are Kenneth Boulding, Professor of Economics and a Program Director of the Institute of Behavioral Science at the University of Colorado, and myself. A case study in the history of decimal time is given by Richard Carrigan, Jr., Physicist and Assistant Head of the Research Division, Fermi National Accelerator Laboratory, and a summary view is presented by David Goldman, Associate Director of Planning, National Measurement Laboratory, National Bureau of Standards.

The reception of the talks at the Houston meeting was mixed. The audience was sparse, reflecting perhaps the conventional view of scientists that metrication is already an accomplished fact for scientific work. Alternatively, the news media was interested in the talks, and their reports generated some correspondence. A particularly cogent letter by Peter Bedford, an English architect, is included in this book.

I gratefully acknowledge the support of my fellow contributors whose timely writings made this work possible.

1.
The Case for Metrication

Louis F. Sokol
President, U.S. Metric Association, Inc.

Perhaps it was my experience as a weather officer during World War II for the U.S. Army Air Corps which brought me to the realization that the world would be a better or at least an easier place to live in if all nations used the same measurement language. For more than a year I was assigned to a joint U.S. Army-Soviet Air Force base in the Ukraine using two spoken languages and two measurement languages in my daily liaison duties. I acquired a first hand knowledge of the simplicity and superiority of metric measures over our customary system of units, if it can be called a measurement system at all.

Following that experience, I vowed to do what I could during my lifetime to encourage the acceptance of the metric system of measurement as the primary measurement language of my country, the United States of America.

Metrication is the commonly used term for changing to the use of the metric system of measurement. Metrication consists of two separate and different functions, although the term can be applied to both of them.

First, metrication refers to the acceptance or implementation of the International System of Units (SI) as a language of measurement. Commonly, the SI is called the modernized metric system. This aspect of metrication is often referred to as a "soft change," since it does not result in a change in the size of objects or things but merely the way in which they are described.

Second, metrication refers to the change in engineering and product standards from those based on inch-pound modules to those based on SI modules and this is often referred to as a "hard change," since this affects the size of things.

From the standpoint of industry, which is leading metrication in the United States, the second aspect of the change is the more important one, and in reality is the reason the change is being made. While the second aspect is more costly, the greatest benefits will be derived from it. The major companies which work on a worldwide basis would like to operate with one set of international engineering standards or common parts, so they can manufacture and service their products in any factory with parts from any source.

In response to some critics of the changeover who say, "Let those companies that want to change make the change but don't force the ordinary people to use metrics," I say that metrication of the big companies also affects their suppliers of parts and services and ultimately their customers, all of whom become involved in using SI metric units. Working people will be more efficient in using metric units on the job if they also use them in sports, hobbies, and consumer-related activities. It is wasteful for a nation to continue to use two major measurement systems, so the sooner complete metrication can be achieved the better it will be for all of us.

The United States is the last of the major nations to have embarked on the road to complete acceptance of the SI as its primary measurement language. To many knowledgeable persons it seems queer that we as a progressive nation should be the last to fully embrace the SI. Much of science has been using metric measures for many years, but their use in industry and commerce has been marginal at best until the last ten years. The fact that the United States has been a world leader in science, industry, and agriculture is due to the initiative and resourcefulness of its people and its abundant natural resources. Progress was achieved in spite of using a complex, poorly related measurement system.

My paper omits most historical and background references, since I assume that most scientists are already generally familiar with these. I will rather dwell on the reasons and need for completing the metrication process which is now well underway in the United States.

The U.S. National Bureau of Standards of the Department of Commerce, in compliance with the provisions of the Metric Study Act passed by the U.S. Congress in 1968, conducted an in-depth three-year study to determine the impact of increasing worldwide use of the metric system on the United States. In July 1971 the secretary of commerce sent his report on the study to the Congress, and included in his recommendations, "that the United States change to the International Metric System deliberately and carefully through a ten-year coordinated national program."[1]

It took four more years before the Congress passed and President Ford signed the Metric Conversion Act of 1975, later known as Public Law 94-168. Unfortunately, the law is weak, establishing no timetable for metric conversion and making adoption voluntary rather than mandatory. The seventeen-member U.S. Metric Board which the law calls for first began functioning in April 1978, and as of this writing has not established an overall policy for metrication in the country.[2]

The law unfortunately also contains the statement, "'metric system of measurement' means the International System of Units as established by the General Conference of Weights and Measures in 1960 and as interpreted or modified for the United States by the secretary of commerce." For political and sociological reasons the secretary of commerce has already modified the SI on the recommendation of metric spokesmen who are promulgating the use of the word 'weight' to mean mass, and the spellings 'meter-liter' instead of the widely used international English language spellings, 'metre-litre.' These spellings were changed in spite of a recommendation to the contrary by Dr. Jean Terrien, Director of the International Bureau of Weights and Measures (BIPM).

During the past two years while industry was moving ahead with metrication, the public sector was floundering because of a lack of a firm commitment from the Congress along with weak leadership from the government. This combined with a misinformed communications media resulted in many anti-metric articles and broadcasts which only contributed to misinforming the general public. It is a human characteristic to resist change; and metrication, as simple as it is, represents a change for most people.

There is only one way that the general public can see the advantages of metrication, and that is by actually using SI metric units. There is no better example than Canada's switch to metric weather reporting and later to metric highway signs. These changes were carried out with determination under the good leadership of Metric Commission Canada. Before the Canadian weather and highway sign change, about 90 percent of Canadian newspapers opposed the change to metric, but after it was accomplished with no dire consequences, 90 percent of the newspapers now favor it.

By contrast, our well-intentioned Federal Highway Administration (FHWA) in April 1977 announced through the medium of a *Federal Register* notice its proposal for metricating the nation's highway speed and informational signs; but it failed in its public relations work before the announcement.[3] Most of us know of the results—the country's two major news reporting organizations flooded the country's newspapers with articles that contained erroneous and misleading information, such as a $100 million cost figure which was not based on facts. This resulted in the FHWA receiving roughly five thousand letters, most of which opposed the change. The letters forced FHWA administrator William M. Cox to promptly rescind the metric highway sign proposal.

How is it that Australia and Canada accomplished the highway sign change at little cost and within a period of one month with no disruption to the movement of highway traffic or inconvenience to the motoring public?

A recent disturbing incident and a total waste of taxpayers' dollars was the recent General Accounting Office (GAO) study supposedly requested by some anti-metric members of the Congress. The GAO study was authorized by the Comptroller General of the United States, and the findings were published in a report to the Congress titled, "Getting a Better Understanding of the Metric System—Implications If Adopted by the United States." The report is biased against metrication and can only serve to slow the conversion process and thereby increase its overall cost to the nation.[4]

Also disconcerting is the attack on the SI by a few individuals from academia who say we should not accept the SI because "their system" is superior. This diversion, while not effective, is difficult to understand, because the SI, while not perfect, is by far the best measurement system that man has been able to devise and promulgate for the needs of science, commerce, and industry. With 95 percent of the world already using the SI, it is much more expedient and practical for the remaining 5 percent to complete the change than for 100 percent of the world to adopt some new, untried esoteric measurement system.

There are four major economic trading groups in the world: the United States, the European Economic Community (EEC), the Council for Mutual Economic Assistance (COMECON), and Japan. Each of the last three has legal directives which require that all documentation originating within or entering any of their member nations use only SI units. This is already affecting U.S. exports to nations in those trading blocs.

Costs of metrication to those companies who are implementing the SI have been truly minimal, because they are following the rule of reason which states that "changes should be made only when they are economically feasible." Generally, this translates into introducing the change in new designs only. Machine tools can work in any measurement system, and some are modified with millimetre readouts for convenience of operation only. General Motors Corporation is perhaps the best example of a company which is changing their operations to the SI. GM has already passed the halfway mark in their metric program, which is being handled routinely just like any other management problem.

Since metrication represents a change, it offers industry an opportunity to review present practices, procedures, product lines, and standards with the possibility of simplifying or rationalizing as many of them as possible. Too often the benefits of metrication are minimized or simply ignored by the uninformed or those opposed to the change. The benefits are indeed substantial as evidenced in countries which have successfully completed the change such as Australia and the Republic of South Africa. The benefits will also continue indefinitely, while costs are strictly one-time factors.

The SI with its seven base, two supplementary, and a host of derived units has a set of sixteen decimally-related prefixes that when added to any SI unit makes a new unit of a more convenient size for specific applications.

The SI has many advantages for engineers. It has clarified the mass-force-weight issue, because it has distinct and separate units for mass and force. Since it is an absolute system in which mass is a base quantity and force a derived one as opposed to our present gravitational system in which force is a base quantity, it does require some effort on the part of engineers to become accustomed to working with it. When using SI units, problems in dynamics dispense with use of the "gravity factor, g" while problems in statics must introduce it.

I pay tribute to the thousands of dedicated and enthusiastic educators who have introduced the teaching of SI metric units in many school districts and some entire states. Not only have these persons found that measurement is more easily taught using the simple, rational SI, they are preparing their students to fill positions in industry which will increasingly require metric knowledge. They are also finding out that children who have learned metric units in school are a vital force in helping to introduce their parents to the world of metrics.

While many scientists have been using the centimeters-grams-seconds system in their professions, they must rapidly swing over to the use of SI units. Many scientists are aware of the superiority of SI units over the customary system, so they should take an active role in promoting their use in all areas of our economy and not only in their professional fields. No new product or idea regardless of its merits is ever accepted without an intensive educational and promotional campaign as any good Madison Avenue advertising executive will attest to.

For this reason I invite all scientists to become members of the U.S. Metric Association and help us in the massive effort of educating and informing United States citizens on the merits and needs for metrication. I am convinced that when the majority of our citizens become aware of the advantages and benefits of going metric, they will not oppose it. If we work hard enough, we could even convince the Archie Bunkers in the country. When it is all accomplished, people will say, "How simple it is, why didn't we do it sooner?"

Notes

1. U.S., Department of Commerce, *A Metric America, A Decision Whose Time Has Come*, National Bureau of Standards Special Publication 345, 1971.

2. Metric Conversion Act of 1975, Public Law 94-168.

3. Metrication of the National Standards for Traffic Control Devices. *Federal Register* 42 (81): 21487, April 27, 1977.

4. U.S., General Accounting Office, *Getting a Better Understanding of the Metric System*, Report to the Congress by the Comptroller General, October 1978.

2.
A Case Study in Metric Conversion: The Optimum Metric Fastener System

Stanley E. Mallen
Metrication Planning Manager, Ford Motor Company

Background

Why haven't we in the United States changed to the metric system years ago? We've always known about the simplicity of the metric system of measurement based on multiples of ten! Essentially, conversion to the metric system has always required extra work and therefore extra cost that was difficult to minimize and justify. Many contributing factors are involved, but Department of Commerce survey results show that the more we know about the metric system, the more we favor it.

In the industrial and commercial field there is a need for creation of improved metric standard product technology with reduced varieties which can potentially reduce final product costs in manufacturing, assembly, sales and service, and subsequently to consumers. This is a major effort and will involve many people. The ANSI (American National Standards Institute)-sponsored special study is a good example of what should be done in other product areas as well.

Engineering standards serve as a dictionary and a recipe book for a technical society and should be recognized as efficient predetermined solutions to potential problems. A complete engineering standard for fasteners normally includes such criteria as dimensions and tolerances, materials, finishes, performance requirements and approved sources of materials.

Used with permission of the Society of Automotive Engineers, Inc. © 1977 SAE. This essay originally appeared as SAE Paper No. 770359, "Metric Fastener Overview."

It is almost impossible to keep a measurement system from influencing standard sizes which normally reflect convenient whole numbers, but we should not be unduly influenced by round numbers. A 2×4, for example, is a convenient reference or name and not an actual lumber measurement. To cover a large range of sizes most effectively, a geometric rather than an arithmetic series should be considered. This geometric progression gives a constant ratio rather than a constant interval between neighboring sizes.

Although larger companies are generally less dependent on industry standard sizes for economical operation, their "specials" are frequently only minor variations from a standard to minimize costs. Small, medium and large companies and consumers benefit from the standardization work being carried on and frequently subsidized by the more conscientious.

It is important to realize that at least four levels of standards evolve and continue to exist at one time. First, a large corporation can specify its needs without much compromise because of its procurement volume. As other companies join in to help create an industry standard, some compromise usually must be accepted. On the national level, less sophisticated industries demand further compromise or a lowering of standards requirements. On the international scene, the less industrialized countries frequently cannot commit to high levels of performance; in some cases even technically competent countries simply refuse to commit to the levels of performance expected in the highly competitive industries in North America.

Some ISO (International Organization for Standardization) standards have been deficient and will probably continue only partially to meet the needs of technically sophisticated users. It behooves all of us, however, to try to overcome these handicaps so that ISO standards come as close as possible to reflecting our common needs.

Optimum standards for such areas as those shown in Figure 1 are required for North America to achieve the maximum benefits of a conversion to the metric system.

Compatibility of standards on a worldwide basis increases the operational efficiency of multinational corporations with a free

**SOME STANDARDS IMPORTANT TO THE
AUTOMOTIVE INDUSTRY WHICH MUST BE DEVELOPED SOON
WITH METRIC MODULE**

BEARINGS	GLASS
THREADED FASTENERS	STEEL AND IRON
	— DIMENSIONS & SHAPES
OTHER FASTENERS	— PHYSICAL PROPERTIES
MECHANICAL DRIVES	NON-FERROUS METAL
— BELTS, CHAINS	— DIMENSIONS & SHAPES
— GEARS	— PHYSICAL PROPERTIES
HYDRAULIC & AIR	RUBBER, TEXTILE & PLASTICS
— FITTINGS	
— EQUIPMENT	PRODUCTION EQUIPMENT
ELECTRICAL	PERISHABLE TOOLS (DRILLS, GAGES, ETC.)
— WIRE & TERMINALS	GENERAL TEST AND PERFORMANCE
— EQUIPMENT	STANDARDS
— LIGHT SWITCHES	

Figure 1.

and clear internal exchange of technical information, raw materials, processing, and common components to increase greater customer satisfaction. The competitive disadvantage of using different standards and measurements is obvious.

Fastener Significance

As a good example of how best to go metric in the long run, let's look at fasteners. Fasteners are nuts, bolts, screws, washers, clips, rivets and a large variety of similar parts used to mechanically join components into assemblies.

In case you are not fully aware of the importance of fasteners in the United States:

1. Annual production is estimated to exceed 200 billion pieces.

2. Annual shipments are worth about $2 billion; the installed value is over $10 billion per year.

3. Almost 2 million different fasteners are produced.

4. About 3 billion pounds of raw material are consumed each year.

Addition of unnecessary parts to a system is expensive to the producer as well as the user. An overproliferation of parts is confusing to designers and greatly complicates the problems of those responsible for procurement and use. There are tremendous opportunities for simplification of fastener sizes, material grades, types, styles, and series.

Fastener Significance at Ford North American Automotive Operations

• At Ford (NAAO), we have used over 12 billion fasteners annually (about 3500 per vehicle).

• NAAO has used up to 11,544 different fasteners per year in 73,000 vehicle applications.

• Approximately $800 million total for fasteners includes installation cost, administration, and approximately $277 million in purchase costs.

Department of Commerce surveys of various manufacturing industries show that total installed costs average approximately 19 times fastener purchase costs.

Figure 2 shows parts released by model years, a key part of our progress report to Ford management. We're proud of our results. We achieved a 38% part number reduction overall in inch fasteners since 1967; and we're striving to continue this improvement. Our program saves millions of dollars annually. At Ford, we consider that fastener complexity costs over $2,000 per part per year, and product reliability improves with component simplification. (The lower part of the graph shows the results of a specific program started in 1971 to track simplification affecting assembly plants only.)

Metrication, however, is temporarily complicating Ford efforts to simplify. The problem will be much less serious than if we had not reduced part number complexity before initiating a transition to the metric system.

ANSI Special Study Committee

In May 1970, the Industrial Fasteners Institute initiated a study to design an Optimum Metric Fastener System (OMFS). Because of the acknowledged potential value of the preliminary results and the interrelated interests of users and others involved in

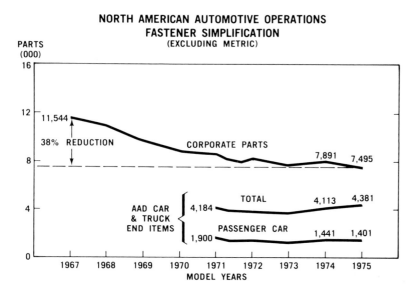

Figure 2.

fastener standards, the ANSI Special Study Committee was established in May 1971 to assure the efficient development of an improved system that will reflect two basic objectives: technical improvement and simplification. In only three years, the committee developed the upgraded system of commercial metric fasteners for consideration by ISO and others.

We explored opportunities to improve the design of mechanical fasteners in any way which would enhance their performance capability. We limited to the fewest possible the number of different sizes, series, types, styles, and grades of fasteners to be recognized as standard. We expected that individual corporations, ANSI, Society of Automotive Engineers (SAE), American Society for Testing and Materials (ASTM), the Canadian Standards Association, ISO, and others would endorse our work by publishing our "Recommendations" as their standards. When the transition has been completed, we expect to save North American industry hundreds of millions of dollars each and every year.

Figure 3 depicts the subcommittee organization. My job as chairman had been simplified tremendously by the cooperation,

leadership, and capability shown by these subcommittee chairmen
and their associates. The Industrial Fasteners Institute (IFI), 1505
E. Ohio Building, Cleveland, Ohio 44114, furnished administrative assistance.

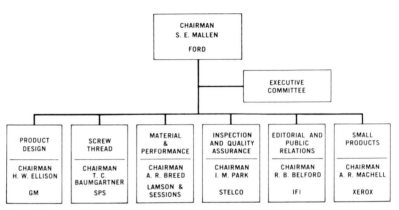

Figure 3.

Committee membership (Figure 4) was broad enough to assure that adequate expertise representing all disciplines was available. The committee included some of the most qualified fastener,
metallurgical, and standards people in North America. Liaison
representatives were also involved.

The ANSI study was actively supported by hundreds of
people working on special assignments and subcommittee efforts.
There were about 175 meetings of the various committees and
task groups.

Figure 5 shows another way of looking at committee com-
position; note the balanced representation of users and producers.
The Special Study Committee developed and recommended, but
did not publish what are normally construed as official standards.
Since we had very qualified representatives and chairmen of affected fastener standards groups on our special study committee,
we anticipated little delay in the translation of our recommendations into standards by these various groups and subsequent recognition as national standards by ANSI and Canadian Standards As-

ANSI OPTIMUM METRIC FASTENER STUDY
– COMMITTEE MEMBERSHIP ROSTER –

CHAIRMAN	— S. E. MALLEN	FORD MOTOR COMPANY
VICE CHAIRMAN	— A. R. BREED	THE LAMSON & SESSIONS CO.
SECRETARY	— C. J. WILSON	INDUSTRIAL FASTENERS INSTITUTE

J. J. Artman	Department of Defense
John W. Axness	Deere & Company
Thomas C. Baumgartner	Standard Pressed Steel Company
D. N. Badgley	Clark Equipment
Richard B. Belford	Industrial Fasteners Institute
D. W. Vial	Chrysler Corporation
Geoffrey Dreger	P. L. Robertson Manufacturing Co., Ltd.
H. William Ellison	General Motors Corporation
N. B. Johnston	International Harvester Co.
Rudy C. Kozlik	National Lock Fasteners — Div. of Keystone Consolidated Industries
Frank V. Kupchak	Westinghouse Electric Corporation
Jack B. Levy	General Electric Company
Arthur R. Machell	Xerox Corporation
John C. McMurray	Russell, Burdsall & Ward Bolt & Nut Corp.
Herman G. Muenchinger	Continental Screw Company
Joseph S. Orlando	ITT Harper Inc.
Ian M. Park	The Steel Company of Canada, Limited
Jack Shugart	Rockford Products Corporation
Arthur G. Strang	National Bureau of Standards
Louis R. Strang	Caterpillar Tractor Co.
William M. Thatcher (Retired)	U.S. Steel Corporation

EUROPEAN LIAISON REPRESENTATIVE
P. J. Gill, G.K.N. (U.K.)

TECHNICAL CONSULTANT
Robert Byrne, U.S. Screw Bureau

AERO SPACE TECHNICAL CONSULTANTS
L. W. Justice, General Electric Co.
B. H. Beal, Grumman Aerospace Corp.

Figure 4.

ANSI OPTIMUM METRIC FASTENER STUDY
– COMMITTEE MEMBERSHIP –

MEMBER AFFILIATION SUMMARY

USER — INDUSTRIAL, AUTO AND AGRICULTURAL	10
GOVERNMENT	2
MANUFACTURER — U.S.	10
— CANADA	2
TOTAL MEMBERS	24
TECHNICAL CONSULTANTS — AEROSPACE AND EUROPE	4
TOTAL	28

Figure 5.

sociation. These recommendations were submitted to ISO Technical Committees TC-1 and TC-2 for adoption as international standards. We expected that the technical merit of the proposals would result in their worldwide adoption as international standards. We were optimistic that an improved ISO commercial metric fastener system would become in time the basis for a single system of preferred standard mechanical fasteners throughout the world.

When we first published the results of the Phase I study in the United States and Europe in 1971, some erroneously assumed that a new thread form and diameter/pitch combinations represented the extent of our interest and that we were trying to create a competing, noninterchangeable system to protect North American fastener producers. Actually, we were developing improvements in a total metric fastener system to accommodate those who might want to convert at some future date to the most appropriate specifications commercially available. In addition to thread studies, the committee investigated such items as:

• New product configurations (for improved performance, identification, etc.).

• Improved material, finish and performance specifications.

• An industry-wide quality assurance program (a "first").

• Improved gaging.

• Simplified variety of fasteners.

Fastener Systems Design Development

Conventional bolt and nut types received priority attention as these represented large volume, structural fasteners. The work involved consideration of the many suggestions and comments received from both North American and overseas contacts.

In the meantime, we developed ten goals for the fastener system:

OMFS Threaded Fastener System Goals

1. The system is to be fully documented.

2. During assembly or installation of a bolt/nut assembly, the

bolt should always experience evident failure before the nut, if overtorqued. (This feature provides immediate failure detection to help insure the reliability of assembled products.)

3. Provide obvious visual and mechanically identifiable characteristics. (This would help minimize possible confusion between metric and other designs in manufacturing, assembly, and maintenance.)

4. Provide rapid visual identification of load carrying capability. (This will reduce the possibility of misapplication of similar-sized product.)

5. Provide positive identification through unique driving means. (This will help eliminate misapplication of product and is particularly important in high-production assembly.)

6. Bolt and nut assembly should optimize flexibility to improve cyclic or fatigue loading characteristics. (Flexible or low-spring-rate fastener assemblies minimize the portion of the external load felt by the bolt and also make the attachment more able to compensate for any changes in joint grip length.)

7. Strength and fatigue parameters shall be at comparable levels. (This goal was aimed at producing a balanced design in the fastener with no gross overdesign of any particular features.)

8. Clearance hole diameters should be associated with each fastener size. (This helps control bearing stresses.)

9. Ultimate designs should conform with good engineering practice which is predicated upon sound economics as related to both product manufacture and application.

10. System should optimize the inherent mechanical fastener property of disassembly for proper maintenance.

Diameter/Pitch Combinations

Figure 6 compares the number and spacing of official ISO inch and metric diameter/pitch combinations commonly available. Small sizes are on the left, large on the right. Top scale is metric, lower scale is inch. Note that metric has 57 and inch has 59 sizes in total. Both the inch and current metric thread series suffer from the standpoint of excessive variety of standard sizes, uneven increments of size, and almost downright redundance when fine versus coarse pitch is considered.

The original IFI proposal in 1970 involved a reduction of standard diameter/pitch combinations to only 25 (up to 100 mm diameter), all in the coarse metric series, but 9 were unique to improve spacing characteristics. The metric coarse series has pitches that are usually between coarse and fine inch threads and therefore is good for standardization. However, in the smaller sizes (3 to 5 mm) the metric coarse thread pitch is finer than inch fine threads which have caused some problems in the past. Because insufficient documentation was available to prove the seriousness of this problem, North America* agreed conditionally to use the ISO coarse thread pitches for these sizes until actual experience justified the effort to revise the pitches.

Other unique proposals to provide better spacing of diameter/pitch standard sizes were dropped in the interest of maintaining consistency with ISO except for one size, the 6.3 mm.

Figure 7 is the same as Figure 6 except that the relative improvement of the proposed 25 diameter/pitch combinations is shown in the middle. The proposal could essentially replace all of the metric and inch sizes. In the mid-range popular sizes, only 6 are proposed where 14 exist in each of the two systems today. The 6.3 mm size is ideally located to preclude the need for two adjacent sizes (6 and 7 mm).

The 6.3 mm size has been approved by ISO as one of its official sizes listed in R261. (The common ISO designator "M" can therefore be used with the 6.3 mm size and other approved sizes instead of the previously proposed "P.") ANSI has also approved the 6.3 mm size for its national standards, and the IFI has listed it as the industry standard. General Motors has listed the 6.3 mm size as standard and has incorporated it in production for the last two years. Unfortunately, some companies have not followed this lead. There is recognition that the 6.3 mm size choice is technically sound to achieve the long-term objective of simplification

---

*"North America" is the collective term for the United States and Canadian fastener industry, both suppliers and users.

COMPARISON OF ISO INCH AND ISO METRIC DIAMETER/PITCH COMBINATIONS

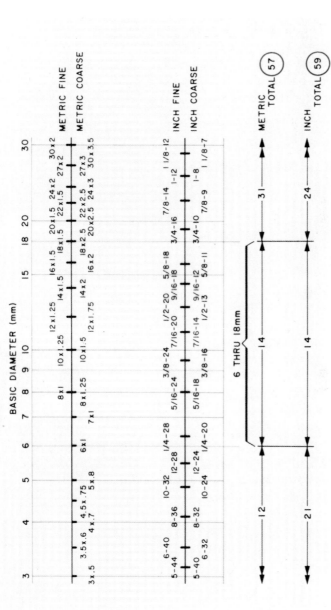

Figure 6.

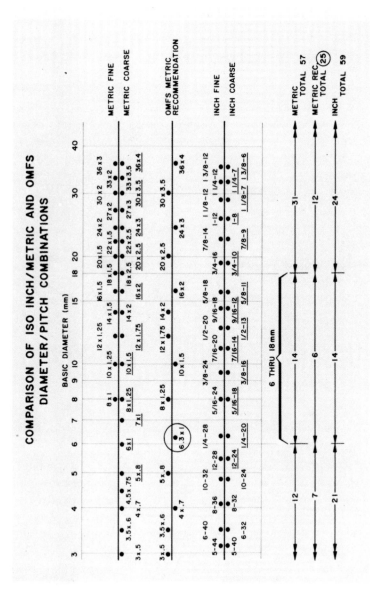

Figure 7.

and that there is no better time to implement it than at the start of North America's metrication program, but supporting effort is inadequate.

Interestingly, the "little guy" (low volume manufacturers, hardware stores, home workshop mechanics, etc.) would probably benefit most from this simplification movement because large volume manufacturers can often justify minor increments in size because of total piece cost savings. The opportunity to continue using the 6 mm size (or any other metric or inch size where it is important to preserve established service interchangeability) is not prohibited by standardizing on a recommended size that is the best for the long run.

Figure 8 compares the ratio of load-carrying capabilities between adjacent sizes. A high ratio is good for simplification and to prevent misuse, but it can be wasteful of material sometimes. The existing ISO metric coarse and ISO inch coarse series are quite erratic; the North American proposal is more consistent but not as good as the 1970 IFI proposal before this proposal was compromised to increase interchangeability with ISO sizes and to provide a 3.5-mm (for comparability with No. 6 in the customary unit size range). Note that the 8 mm is 82% stronger than the 6 mm.

Figure 9 shows the North American recommended industry standard commercial metric diameter/pitch combinations.

The curves in Figure 10 show the cumulative percentage of fasteners used by size and strength grade. This distribution data base was used to evaluate the efficiency of alternative selections of diameter/pitch combinations.

Figure 11 shows the relative efficiency of material usage (and, therefore, cost) for some selected diameter/pitch combinations using a computer simulation technique. We wanted to expose the penalties incurred by having to use fasteners larger than necessary when the next smaller size does not meet the requirements. As you might guess, a large number of diameter/pitch combinations should be more efficient in material usage than a small number of choices; however, adequate spacing is also important to prevent

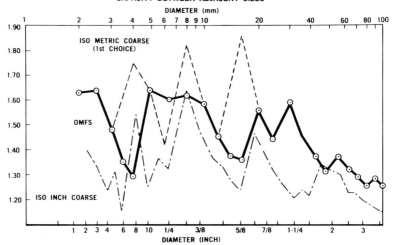

Figure 8.

M1.6x0.35	M20x2.5
M2x0.4	M24x3
M2.5x0.45	M30x3.5
M3x0.5	M36x4
M3.5x0.6	M42x4.5
M4x0.7	M48x5
M5x0.8	M56x5.5
M6.3x1.0	M64x6
M8x1.25	M72x6
M10x1.5	M80x6
M12x1.75	M90x6
M14x2	M100x6
M16x2	

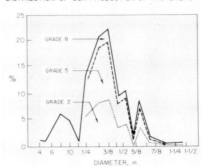

DISTRIBUTION OF USA PRODUCTION OF FASTENERS

Figure 9.   Figure 10.

mixing and to minimize the substantial indirect costs of proliferation.

COMPUTER SIMULATION RESULTS
SIZE RANGE 5 TO 36mm (NO. 6 TO 1½")

	ISO METRIC R 261 FIRST CHOICE	ORIGINAL IFI PROPOSAL	OMFS PROPOSAL	ISO INCH COARSE
NUMBER OF SIZES	10	10	11	14
RELATIVE COST	104.9	103.5	100.0	100.0

Figure 11.

Thread Forms

The ISO metric thread form as described in ISO 68 is similar to our Unified Inch thread profile standard since 1948. However, ISO metric external threads could have a flat, sharp cornered root. North American practice has been to require a radius root to improve fatigue and static strength and improve the life of thread rolling dies. The standard UNR* thread has a minimum root radius of 0.108 times the thread pitch; for aerospace UNJ* the minimum ratio is 0.15P, and we had proposed a minimum radius of 0.14P to achieve commonality for aerospace as well as commercial applications.

Basic thread form depth for metric and inch fasteners is 0.54127P; for UNJ (aerospace) it is 0.48. OMFS had proposed a compromise depth of 0.5P to further help achieve worldwide commonality of threads.

---

*UNJ and UNR are the official designations for the inch-based aerospace and commercial thread forms in common use.

To explain evolutionary developments, Figure 12 shows the previously proposed external thread form and the originally proposed truncation of the internal thread to assure compatibility. We subsequently determined that ISO metric internal threads were functionally interchangeable with the radius root recommended for external threads and did not have to change.

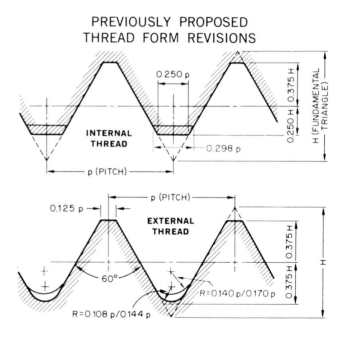

Figure 12.

To support the 0.14P radius proposal, we investigated the difference between possible and statistically probable thread interference or "encroachment" (Figure 13). Figure 14 shows that the probability of actual interference (based on Monte Carlo computer or random assembly simulation results for 4000 measured nuts and bolts) was zero or negligible for popular sizes even though interference theoretically could have occurred. Even with encroachment in the high pitch sizes, the amount of interference was considered tolerable. After reaching the compromise described below, even theoretical interference was eliminated.

In developing the basic profile or maximum material limits for internal and external threads, we questioned use of theoretical triangles and pitch diameter. Our studies have shown that they could be omitted.

Figure 15 shows the new and simple basic profile. The basic height of 0.5P is no longer recommended because the aerospace industry chose not to accept the compromises. For this reason, we reverted to the previous standard height of 0.54127P and adopted a more acceptable external thread root radius of 0.125P as the minimum. ISO TC-2 adopted the compromise and will require this minimum root radius for externally threaded parts with specified tensile strength of 830 megapascals (120,000 psi or SAE grade 5) or higher.

There will be only one tolerance class (6H) for internal threads and two classes for external threads: 6g for general purpose and 5g6g for closer tolerance threads.

In summary, ISO and modified ISO (OMFS) threads are now identical.

Thread Gaging

It is important to have gaging practice which meets design requirements. Today's methods do not provide this assurance, as shown in Figure 16. A thread form with an error in flank angle and major diameter could offer the "definite drag" currently required with present NO-GO gages.

Figure 17 shows the new double NOT-GO gaging of internal threads recommended by OMFS for sophisticated requirements when similar gages are used to set up equipment and for process control. The two NOT-GO gages assure minimum material in the product. Note that, should the flank not reach the boundary, at least one NOT-GO gage would reject the part.

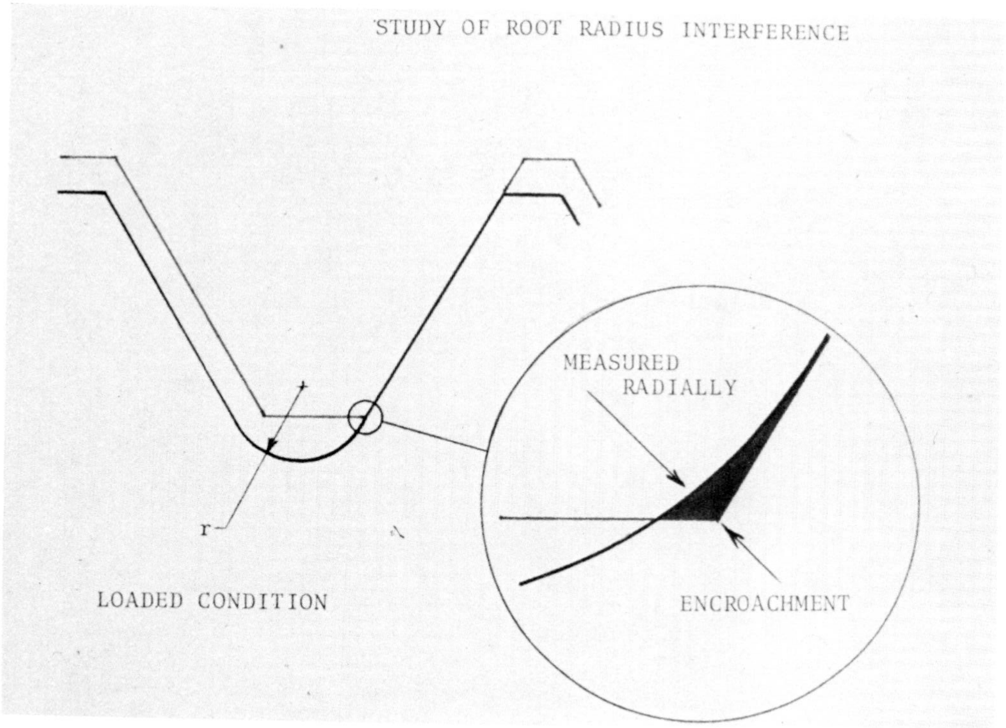

STUDY OF ROOT RADIUS INTERFERENCE

MEASURED RADIALLY

ENCROACHMENT

LOADED CONDITION

r

Figure 13.

LIMIT
CONDITIONS   PROBABILITY OF INTERFERENCE

PITCH mm	NO INTRUSION	0.025 mm ENCROACHMENT	0.050 mm ENCROACHMENT
.25			
.30			
.35			
.40			
.45			
.50			
.55			
.60			
.65			
.70			
.75			
.80	.1		
.90	.1		
1.00	.2		
1.25	.3		
1.50	.7		
1.75	.9		
2.00	.9		
2.50	1.8	.3	
3.00	2.3	.4	.1
3.50	3.2	.9	.2

Figure 14.

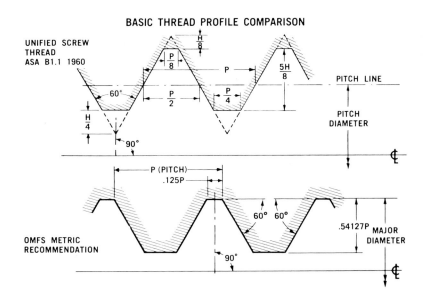

Figure 15.

Incidentally, the GO ring or plug gaging remains unchanged; it continues to check "maximum material limit" for the product. Conventional "NO-GO" gages are recommended for general purpose part production.

Prototypes of the new gaging technique have demonstrated the feasibility and value of the double NOT-GO approach. Early in the development of the concept, a plug gage was made with the two NOT-GO probes for the ¼-20 size. A series of 9 tapped holes had been prepared, all toward minimum material condition, 3 with a 60° thread, 3 with a 48° thread, and 3 with a 66° thread. All 9 holes accepted the standard GO thread plug gage and refused the standard HI thread plug gage. Thus all 9 tapped holes were "acceptable." Using a new double NOT-GO probe technique, the 3 holes with 60° threads were acceptable; that is, both NOT-GO gages were refused. The other holes with 48° and 66° threads were rejected; that is, each permitted entry of one of the NOT-GO probes.

TWO EXAMPLES OF IMPROPER INTERNAL THREAD FORM
NOT DETECTABLE BY PRESENT GAGING TECHNIQUES

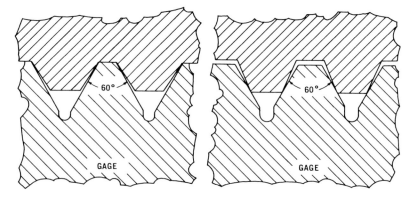

Figure 16.

PROPOSED INTERNAL THREAD GAGING FOR
OPTIMUM METRIC FASTENER SYSTEM

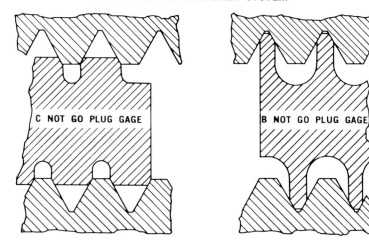

Figure 17.

Figure 18 shows the proposed external thread gaging system using NOT-GO gages.

PROPOSED EXTERNAL THREAD GAGING FOR OPTIMUM METRIC FASTENER SYSTEM

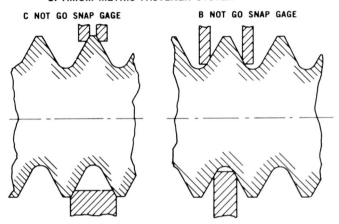

Figure 18.

For checking internal and external threads, more sophisticated "boundary profiles for gaging" (Figures 18a and 18b) have been defined to replace the product "thread limits of size" concept.

Bolt Head Design

The following 15 basic criteria were initially established to evaluate the adequacy of any new bolt head design proposed as an "optimum" configuration:

1. The design of the head should provide a minimum head strength of 110% of maximum bolt-tensile strength.

2. Head material content should be minimized.

3. Spring rate of the bolt-nut assembly should be minimized.

4. The driving configuration of the head shall withstand a torque of 300% of the torsional limit of the threaded portion of the shank without detrimental deformation.

5. The head should lend itself to economical manufacture.

6. The outside diameter of the head/driver should be minimized.

BOUNDARY PROFILES FOR GAGING INTERNAL THREADS

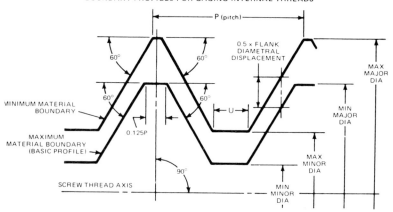

Figure 18a.

BOUNDARY PROFILES FOR GAGING EXTERNAL THREADS

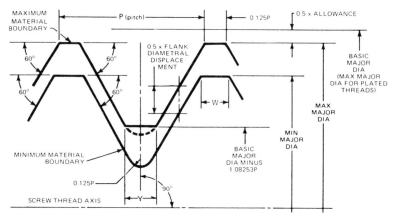

Figure 18b.

7. The driving tool costs should be minimized.

8. The configuration of the fastener should be unique for system identification purposes. Additional information should be provided for strength capability and manufacturer.

9. The head height should be minimized.

10. Assuming uniform loading under the head, the bearing stress should not exceed 345 MPa at ultimate tensile strength.

11. Design should incorporate the maximum fillet radius that is practical—as generous as possible under the head at the junction between shank and head.

12. Clearance holes should be specified with initial consideration given to existing ISO clearance hole sizes.

13. The head design should be compatible with high-speed assembly.

14. Fatigue failure should occur in the threaded section and not in the head.

15. It is assumed for design purposes that the shank will experience a torque equal to the torsional failure torque of the shank as predicted by the distortion-energy theory when the tensile load in the shank is equal to 80% of the ultimate axial tensile strength.

Following the establishment of the fastener system goals and the bolt head design criteria, a task group (headed by Mr. Jack Shugart of Rockford Products, Inc., Rockford, Illinois) was organized to study and make recommendations concerning the bolt head and shank design.

Since 11 basic unique configurations and some variations had been formally considered in addition to the many so-called "standard" fastener shapes currently available in inch and metric systems, some means of analyzing the candidates systematically and fairly was required. Dr. Romualdas Kasuba (Cleveland State University) was retained to perform this preliminary analysis using computer programs to assist in the design evaluation. The bolt stress analysis was followed by extensive manufacturing and assembly feasibility studies.

The "optimum" new bolt head configuration had been developed in 5 to 20 mm diameter and some initial samples are shown in Figure 19.

NEW OPTIMUM METRIC FASTENER SYSTEM (OMFS) being proposed for adoption by U.S. industry is illustrated by these 12-spline flange screws manufactured for test purposes. From left to right, their diameters are 24, 16, 12, 12, and 8 mm (millimetres), respectively.

Figure 19.

The 12 spline flange screw (not 12 point) has characteristics superior to any current fastener (metric identification, wrenchability, material utilization, and manufacturing flexibility). Prototype parts in six sizes had been manufactured by twelve companies for physical test purposes (Figure 20). The new design had been developed cooperatively with Great Britain, Sweden, and Germany. Very little production of this design has resulted, however, primarily because of higher than expected pricing.

SOME FASTENER PRODUCERS ACTIVELY DEVELOPING PRODUCTION CAPABILITY FOR 12 SPLINE SCREWS

COMPANY
BULTEN-KANTHAL (SWEDEN)
CONTINENTAL SCREW
FEDERAL SCREW WORKS
GKN (GREAT BRITAIN)
ITT HARPER
KELLERMAN (GERMANY)
LAMSON & SESSIONS
MODULUS CORP., SCREW & BOLT DIVISION
ROCKFORD PRODUCTS
RUSSELL, BURDSALL & WARD
STANDARD PRESSED STEEL
STEEL CO. OF CANADA
TOWNE ROBINSON
NATIONAL LOCK (KEY CONSOLIDATED INDUSTRIES)

Figure 20.

A series of hex-head screws and bolts reflecting efficient use of material has been developed (for those who do not need the flanged, 12 spline design) to save 50 tons of material and millions of dollars annually. The ANSI/OMFS committee reduced ISO hexagon width-across-flats (WAF) to standard ratios as follows:

Hex Head Width Dimensions

Nominal Size	Width Across Flats (mm)	
	ISO R272	OMFS/ANSI/IFI
M 10	17	15
M 12	19	18
M 14	22	21

The next revision of ISO R272 should list those new sizes so that commercial practice can influence resolution of this choice; there is very little objection to changing M 12 and M 14.

Hex-head height dimensions are being changed as follows:

Comparison of Hex Head Height Dimensions

Nominal Screw Size, mm	Height of Head, mm	
	ISO R272	ANSI/IFI & DIS 4014, 4015
8	5.6	5.3
10	7.0	6.4
12	8.0	7.5
14	9.0	8.8
20	13.0	12.5
30	19.0	18.7
36	23.0	22.5
90	57.0	56.0
100	63.0	62.0

Hex flange screw standards are being developed by ISO and will probably reflect the current practice (initiated by Ford Motor Company and rapidly being accepted as an industry standard) of using the next smaller hex size of the regular hex series, as follows:

Nominal Size	Width Across Flats (mm)	
	OMFS/ANSI/IFI	Hex Flange Screw
M 10	15	13
M 12	18	15
M 14	21	18
M 16	24	21

(See Mr. J. F. Nagy's SAE paper No. 770422, "Hex Flange Bolt Head Weight Reduction Design Criteria" and Figure 22.)

The series will help assure standardization on the 15 mm
WAF to reduce the number of sockets and wrenches required. A
new series of 12 spline socket wrenches was developed which in
some cases can also drive a hex head bolt.

The fillet under the bolt head is a simple radius in ISO standards, but a parabolic or compound radius fillet and longer transition zone is the recommended practice for North America to reduce the maximum stress concentration. (The undercut radius
fillet for flange screws can be specified by the customer.)

Materials and Performance

To get an idea of why we're concerned about technically infe-
rior standards, take a look at Figure 21 comparing ISO and
SAE/ANSI physical properties for ostensibly equivalent material
strength grades. If we were to adopt existing ISO strength standards, we would actually downgrade some of our important standards. The new class 9.8 is recommended as the new metric replacement for the popular SAE Grade 5 and has been approved by ISO.

COMPARISON OF ISO METRIC AND SAE STRENGTH
SPECIFICATIONS FOR AN EQUIVALENT MATERIAL GRADE
(MEDIUM CARBON STEEL, QUENCHED AND TEMPERED)

GRADE	SPECIFIED MINIMUM TENSILE STRENGTH MPa	RELATIVE AVERAGE COST PER UNIT OF CLAMP FORCE $ / MN	
ISO CLASS 8.8	800 (116,000 PSI)	$.58	
SAE GRADE 5	830 (120,000 PSI)	$.56	14% DIFFERENCE*
OMFS PROPOSAL (ISO CLASS 9.8)	900 (130,500 PSI)	$.51	

Figure 21.

BOLT-HEXAGON FLANGE HEAD
THREAD CLASS 6g
PROPERTY CLASS 9.8
PITCH DIAMETER BODY
CATEGORY 01.02

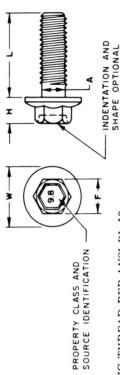

PROPERTY CLASS AND
SOURCE IDENTIFICATION

INDENTATION AND
SHAPE OPTIONAL

METRIC THREAD PER ANSI B1.13

Dimensions in mm

Nominal Size		M5	M6	M8	M10	M12
Pitch mm/Thd.		.80	1.00	1.25	1.50	1.75
A	Body Diameter	4.46	5.33	7.16	9.00	10.83
		4.36	5.21	7.04	8.86	10.67
F	Hex Width-Ref.	7.0	8.0	10.0	13.0	15.0
H	Head Height-Ref.	5.2	6.3	7.6	8.6	10.8
W	Flange O.D.-Ref.	10.8	13.0	16.2	20.0	24.0

Figure 22.

An OMFS strength grading system for ferrous fasteners comprising 7 property classes (dropping 6 listed by ISO) has been designed and adopted by ANSI and IFI. Only 3 property classes are standard for fasteners smaller than 5 mm and only 4 for sizes larger than 24 mm. (Figure 23.)

In the inch series, over 80 strength grades are available for ferrous fasteners.

All properties are either equivalent or improved over those now standard for inch series fasteners and/or ISO metric fasteners. The upgrading reflects what is commercially available without significant change in processing and will frequently permit the use of smaller (lower cost) metric fasteners than previously required.

A single system of head marking symbols to designate strength grades of ferrous fasteners has been accepted internationally. These markings have not been used by North American manufacturers on inch fasteners; consequently, they will help identify metric products. (Figure 22.) A blue color has been used selectively by some as additional metric fastener identification where risk of mixing with inch products nearby is a significant factor. The practice should be very restricted because of additional proliferation, cost, possible adverse effect on torque/tension relationships, appearance, and because in the long run practically all fasteners will be metric.

Strength grade systems for stainless steels and nonferrous fasteners have been established in metric for the first time. (See Mr. J. S. Orlando's SAE paper No. 770419, "Corrosion Resistant Steel and Nonferrous Materials for Metric Fasteners.")

Nuts

Standards for free running hex, hex slotted, hex flange nuts and prevailing torque hex and hex flange nuts have been developed. Two styles (thickness) of hex nuts have been designed for the three popular property classes. (Figure 25.) Property class 8 hex nuts have been designed to incorporate significantly greater resistance to thread stripping, a very undesirable mode of failure.

**PROPERTY CLASSES FOR IFI
METRIC BOLTS, SCREWS, AND STUDS**

Property class	SAE grade [1]	Nominal diam (mm)	Proof load (MPa)	Tensile strength (min. MPa)	Rockwell hardness min.	max.
4.6	1	M5 thru M36	225	400	B67	B100
4.8	—	M1.6 thru M16	310	420	B71	B100
5.8	2	M5 thru M24	380	520	B82	B100
8.8	5	M16 thru M36	600	830	C24	C34
9.8[2]	—	M1.6 thru M16	650	900	C27	C36
10.9	8	M5 thru M36	830	1040	C33	C39
12.9	—	M1.6 thru M36	970	1220	C39	C44

[1] To be used for guidance purposes only in selecting metric property classes.

[2] This class is actually 9% stronger than SAE Grade 5 and ASTM A449.

Figure 23.

How to interpret property class numbers

Digits to the left of the decimal point represent approximately 1/100 of the minimum tensile strength in megapascals (MPa).

EXAMPLE:

In property class 5.8, the 5 represents the 1/100 approximation of the 520 specified minimum tensile strength.

Digits to the right of the decimal point represent approximately 1/10 of the ratio, in percent, between minimum yield stress and minimum tensile stress.

Figure 24.

(Figure 26.) (See Mr. E. M. Alexander's SAE paper No. 770420, "Analysis and Design of Threaded Assemblies.")

A number stamped on the nut is the normal metric identifier.

Small Screws

Five separate metric standards have been developed for tapping screws, thread rolling screws, self-drilling screws, machine screws, and screw and washer assemblies.

Five popular head styles for tapping and machine screws are standard vs. 15 for inch. Only 7 different thread types for thread

PROPERTY CLASSES AND STYLES OF IFI METRIC HEX NUTS

Property class	Description	Style°	Nominal size (mm)	Proof load stress (MPa)	Hardness (Rockwell)	Mating bolt/ screw class
5	Low carbon steel Nonheat treated	1	M5 thru M36	570	C 30 max	5.8, 4.8, 4.6
9	Low carbon steel Nonheat treated	2	M1.6 thru M4 M5 thru M16 M20 thru M36	900 990 910	C 30 max	9.8, 8.8, 5.8, 4.8
10	Medium carbon steel Heat treated	1	M5 thru M36	1040	C 26/36	10.9, 9.8, 8.8

*Style is based on thickness.
Style 2 is thicker than Style 1

Figure 25.

TYPICAL BOLT & NUT ASSEMBLY CHARACTERISTICS

(COARSE THREAD)

SAE MATERIAL GRADE OR ISO METRIC PROPERTY CLASS	TENSILE STRENGTH		RATIO OF NUT THICKNESS TO NOMINAL DIA.	APPROX. THREAD STRIPPING PROBABILITY (IF OVERTORQUED)
	PSI	MPa		
SAE GRADE 5	120,000	830	.85	1%
CLASS 8	116,000	800	.8 (PRESENT)	UP TO 40% (DEPENDS ON SIZE)
			.9 (PROPOSED)	1%
CLASS 9	130,500	900	1.0	1%

Figure 26.

forming and thread-cutting tapping screws are standard vs. 13 previously; diameter/pitch combinations agree with ISO standards. (Essentially a "soft conversion" of present inch products, commonly used around the world.) Other tapping screws and all machine screws are hard metric sizes.

In order to have one worldwide standard for the conical bearing surface, flat and oval countersunk head machine and tapping screws will have a 90° head instead of the 80° angle standard for inch series. Phillips and Pozidriv cross driving recesses are listed as standard without preference, in common with ISO practice. It is unfortunate that the slight economic difference is enough to continue this proliferation adversely affecting producers, assemblers and ultimate consumers.

(See SAE paper No. 770423, "Design of Heads for Metric Machine and Tapping Screws" by Ray Ollis, Jr.)

Blind Rivets

Since blind rivet production and use around the world conforms with inch series sizes, metric blind rivet standards were developed by "soft conversion."

Round Head, Square Neck Bolts

One style of metric round head, square neck carriage bolt has been developed as standard compared with three now available in inch.

Clearance Holes

Although rational standards for clearance holes (the sizes of holes in parts to be joined by fasteners) are not defined in the inch system, metric clearance hole standards have been established.

Washers

The SAE Fasteners Committee is finalizing metric flat washer standards in narrow, regular and wide series and is developing metric conical washer standards.

Dryseal Pipe Threads

ANSI has published a metric standard (B-1.20.4-1976) that is a soft conversion of the Inch Dryseal Pipe Thread Standard (B-1.20.3-1976) used around the world.

Inspection and Quality Assurance

Inspection constitutes an increasingly significant portion of total manufacturing cost, and if not carried out effectively, is disruptive to operations. Because of the high costs associated with the multiplicity of quality assurance plans now specified, and since ISO proposed drafts are unacceptable to North America, the ANSI/OMFS committee developed a recommended plan for worldwide use. (OMFS-12)

North American practice is frequently based on the concept of not knowingly accepting defective parts (zero-defectives in the inspection sample) and lot control. ISO recommends acceptance of shipments when the inspection sample discovers relatively high "acceptable number of defectives" and seriously restricts the number of fastener characteristics which will be checked at final inspection.

Lot control is not favored by fastener producers outside North America.

The OMFS-12 recommendation allows flexibility to cover four types of fasteners or assurance requirements:

1. General purpose fasteners (hardware store types).
2. Fasteners for high volume machine assembly.
3. Fasteners for special purpose applications requiring in-process control.
4. Fasteners for highly specialized engineered applications produced consecutively from a single mill heat with in-process control and lot traceability.

Technically sound, practical for producers, and adequately protective of fastener users, the plan was offered in October 1974, a timely opportunity for industry to standardize on a quality assurance approach. This would be an "industry first."

SAE is considering the recommendation, but chances of establishing a satisfactory acceptance procedure in ISO are slim.

Administrative and Technical Support

The Industrial Fasteners Institute deserves special commendation for the foresight shown in launching the ANSI/OMFS study and in providing significant administrative, financial and technical support. It has been estimated that over $1 million of research and engineering effort was devoted to the study by IFI, its member companies, and other organizations represented on the study committee. The IFI provided administrative coordination of all official committee actions including technical editing and publication of committee meeting notifications, minutes, agenda and progress reports. They provided other organizations with information on OMFS activities; there have been press releases, multi-city North American presentations, and technical articles published. We also reported every few months on our activities to our European friends and have sent representatives annually overseas to discuss our progress. We published annual reports and special technical reports. Especially recommended is the publication "Transactions of Technical Conference on Metric Mechanical Fasteners" (March 1975) available from ANSI, 1430 Broadway, New York, NY 10018.

The IFI published individual "OMFS Recommendations" as our work was being completed during the 1972-75 period, so that industry, associations and others could start developing their metric fastener standards, if necessary.

The ANSI/OMFS committee terminated its activities on July 10, 1975 and the Industrial Fasteners Institute then prepared and published in March 1976 the most complete consolidation of the latest metric fastener technology in an extensive volume entitled *Metric Fastener Standards*. Figure 27, a reproduction of the cover of *Assembly Engineering* magazine not only shows the handbook but also pays tribute to Richard B. Belford, the very able Technical Director of IFI. *Metric Fastener Standards* is the first book published in English setting forth a system of metric mechanical fasteners and contains 32 separate standards. ISO fastener standards are feature-oriented so that separate documents must be compiled to completely define a specific fastener. In addition, ISO standards must reflect extensive technological and proliferation compromises to satisfy the interests of many countries. Until ANSI, SAE, ASTM and American Society of Mechanical Engineers (ASME) official standards are published, the IFI handbook is a very important document. In addition to the subjects previously

Figure 27.

covered in this paper, *Metric Fastener Standards* covers studs; test methods; surface discontinuities; bent bolts; lock screws; high strength structural bolts, nuts and washers; socket screws; spring pins; tooth lock washers; rules for the use of SI metric units; a list of related ISO Standards and Recommendations; and a glossary of terms.

The SAE Fasteners Committee in addition to developing official metric standards in most of the areas mentioned (and others) is also developing an inventory of fastener standards. The E21/E25 Joint SAE Committee has published 3 metric documents (1337, 1338 and 1370–"J" thread).

The ANSI B-1 (Screw Threads) and B-18 (Fasteners) Committees are developing or have published metric standards as follows:

- Screws, Hexagon Socket Head Shoulder (Metric Series), ANSI B18.3.3-1979
- Screws, Metric Formed Hex, ANSI B18.2.3.2M-1979
- Screws, Metric Heavy Hex, ANSI B18.2.3.3M-1979
- Screws, Metric Hex Cap, ANSI B18.2.3.1M-1979
- Screws, Metric Hex Flange, ANSI B18.2.3.4M-1979
- Screws, Metric Hex Lag, ANSI B18.2.3.8M-1979
- Screws, Socket Head Cap (Metric Series), ANSI B18.3.1-1978
- Bolts, Metric Heavy Hex, ANSI B18.2.3.6M-1979
- Bolts, Metric Heavy Hex Structural, ANSI B18.2.3.7M-1979
- Bolts, Metric Hex, ANSI B18.2.3.5M-1979
- Keys and Bits, Hexagon (Metric Series), ANSI B18.3.2M-1979
- Metric Screw Threads–M Profile Standard is ANSI B1.13M-1979
- Gaging Systems for Screw Thread Dimensional Acceptability Standard is ANSI B1.3-1979
- Metric Screw Threads, American Gaging Practice Standard is ANSI B1.16-1972
- Screw Threads-Metric Series MJ Profile Standard is ANSI B1.21-M-1978

- "Gages and Gaging Practice for Metric J Screw Threads" Standard is ANSI B1.22-1978

Note: ANSI B-1 has agreed that:
(a) References to referee gaging methods will be eliminated from B-1 product documents.
(b) The level of dimensional acceptability will be determined by the threaded product application and specified by product standards or by procurement drawings or documents.
(c) A catalog of gaging systems will be published without any indication as to preference for a given application.

Hardware Availability and Selection

There is a general misconception regarding the availability of standard fasteners, whether inch or metric. "Standard" does not necessarily mean "on the shelf." "Standard" means that the product is listed and described in a reputable document. Since standards are published by companies, industry associations, technical societies, national standards groups, and ISO, these documents will seldom agree completely because of changing technology, different publication dates, compromises required to achieve agreement, honest differences of opinion regarding the importance of various product characteristics, resistance to change, etc. Stocking "standard" products in anticipation of need (except in the case of hardware store or general purpose fasteners) can be expensive because of the variety of specifications used by industrial and commercial customers. Fastener producers are very willing to produce what customers want, but availability of small quantities for prototype or experimental purposes will always be a problem, but one of economics and lead time rather than technology.

Selection of hardware for engineering release, procurement and production use requires the resolution of a dilemma. Since we all would like to use internationally approved metric standard fasteners, what should we do if the official documents do not yet reflect the most appropriate product considerations? What if compromises to be reached in large groups such as at the international level are not commercially and technologically satisfactory to the engineer, his management and his company's customers?

Just as before in the use of inch type fastener standards, we have to use our own best judgment regarding the practical alternatives. We should select the soundest technological specification details in engineering a product and assume that others will do the same so that popular consensus is reached in the competitive market. Proven standards will then evolve to confirm good commercial practice. Compliance with standards is not mandatory unless required by law, but deviation from past practice should be tolerated and encouraged if progress is to be achieved.

Incidentally, when people ask about ISO "resistance" to change, I remind them that if the other nations of the world were adopting the customary unit measurement system, many in North America would probably be resisting the various improvements that could be achieved.

The IFI *Metric Fastener Standards* handbook is the best consolidation of metric fastener information available today and it will probably continue to be for some time. It should be used as the starting point for most industrial fastener engineering work.

A personal regret is that the OMFS Committee was unable to agree on a standard catalog part numbering system so that popular basic parts could be ordered anywhere in the world using part numbers only. Some work was done to develop a simple part numbering system with significant numbers to indicate type of part, specific variation, nominal diameter and pitch, length, material strength and finish.

It became apparent that large companies having need for very sophisticated systems would not change, so until small and medium size companies show some interest no action is being taken.

Conclusion

The chance to improve significantly something as basic as our system of measurement and many of our fundamental engineering standards seldom comes in one's lifetime, and we must exploit this golden opportunity. We have alerted responsible people to the value of a coordinated, planned conversion to the metric system based on: 1. improved worldwide standards and product "rationalization," 2. greater participation in developing metric worldwide standards reflecting superior design and manufacturing technology, 3. introduction of these upgraded metric standards into product development and production with enough coordinated corporate planning that other possible improvements are also incorporated.

To illustrate the value of a coordinated approach to metrication, let's look at Ford's fastener simplification potential. Figure 28 shows Ford's past progress and my prediction of future progress without a formal worldwide fastener standardization program. I hasten to add that we do have a coordinated voluntary program on each side of the Atlantic. Although Ford is a worldwide company, we provide our overseas affiliates with a fair amount of latitude regarding their product decisions. Because of measurement unit differences, different standards and fasteners are used. As we adopt one measurement system and as world trade increases, however, interchangeability becomes more attractive.

As a company we could have standardized on the inch (or soft conversion equivalent), the present ISO metric, or an improved version of both. Since metrication is becoming increasingly popular, some kind of metric system was the only practical candidate. If North America is to make a basic change of this magnitude, however, it ought to be to the best available system to preclude the expense of a second change at a later date.

While transition to the metric system will cause temporary problems, if we recognize the long range advantages of simplified but technically superior metric specifications (and develop long-term financial policies to encourage their use) proliferation of varieties will be controlled and long-term benefits will accrue to everyone. Also, if future nonstandard, low volume, special products are priced without adequate recognition of intangible, administrative, and other related costs, there will be little incentive on the part of users to order the potentially large volume industry standard optimized metric parts.

Figure 29 shows Ford's past accomplishments and my prediction of future progress when the coordinated worldwide fastener

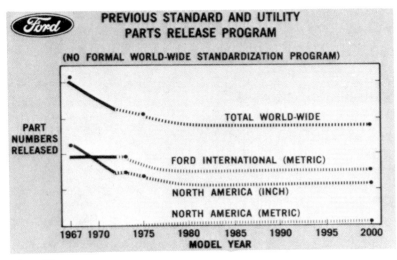

Figure 28.

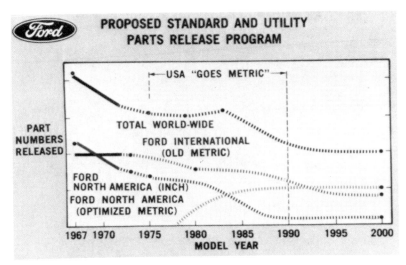

Figure 29.

standardization program is implemented. The potential level shown in Figure 28 is 40% higher than the level shown here.

Optimized metric fasteners can be the basis for direct and administrative savings of hundreds of millions of dollars, not just one time, but repeated each and every year.

3.
Numbers and Measurement on a Human Scale

Kenneth Boulding
Professor of Economics and a Program Director,
Institute of Behavioral Science,
University of Colorado, Boulder

The controversy regarding the "scientific" character of the metric system and its possible alternatives is part of a larger and very fundamental problem in epistemology. This might be called the "number question." This is the question as to the role of numbers, whether they result from counting or from measurement, in the learning process of the human mind, that is, the process by which it acquires knowledge. The human nervous system has an extraordinary capacity for taking structure, the physical basis of which is still largely unknown but which reveals itself in language and communication of all kinds—verbal, mathematical, numerical, symbolic, oral, visual, even tactile. These structures have a remarkable property of being able to transfer themselves from one mind into another. These structures exhibit some kind of mapping onto an external reality, with at least an approximate one-to-one correspondence.

We can distinguish a number of different grades of this mapping. "Truth" consists of those internal structures of the mind which are accurate maps of an external reality. The accuracy may be a matter of degree and is probably never perfect, just as a map is never a perfect representation of the area that is mapped. Error is the quality which a structure in the mind has of not corresponding to an external reality. Hypothetically, therefore, we could take any particular structure or image in the human mind and rank it on some kind of truth-error scale. The valuation system of the human mind does this constantly. We run into something of a paradox here: where we evaluate a particular image as being in error, we tend to modify it towards what we regard as truth.

We also, however, have another dimension of valuation, which is uncertainty or ignorance. If I ask myself what is the state capitol of Alabama, I may simply come back with the answer "I don't know." Or I may say, "Maybe it is Birmingham, but I am not sure." And we may have some kind of valuation of the probability or likelihood of the truth of these answers, even if we do not put a number on it. Unlike an evaluation of error, an evaluation of ignorance may be quite stable, although, if the sense of curiosity is strong, an evaluation of ignorance may lead to activity designed to reduce it. In the above case I may go to an authoritative source like an atlas and simply look up the state capitol of Alabama. As a matter of fact, I just did so and discovered that the state capitol was not Birmingham but Montgomery. That is, I conducted a test which substituted a high belief in knowledge for ignorance and possible error. My belief of knowledge in this case I must confess is based on authority. My belief in the authority of atlases is very strong; they have very rarely let me down. If I am suspicious of the authority, however, I could conduct an experiential test, go down to Montgomery and see if it has the state capitol. Then I am sure that my confidence in this knowledge would be 100 percent instead of the 99.999999 percent it is now.

This raises, however, a very tricky problem in epistemology, first brought to prominence by David Hume. This is that we can never directly compare a map in our minds with the reality outside it. We can only compare a map with another map. If I had gone down to Montgomery, Alabama, and discovered everybody I talked to denied it was the state capitol, if I could find no capitol building, the image in my mind, particularly the reliability of atlases, would change. I would compare the old image with what I regarded as a more reliable new one, and would have discarded the old image as a result, including my images of the reliability of atlases as authorities.

The scientific community is a subculture within which there is a very subtle and complex interaction between authority, the testing of authority, and personal testing. The image of most scientists of their own science and of all others is derived primarily from authority. The authority of textbooks, of teachers, of periodic tables on the walls of chemistry lecture rooms, authority reinforced by success in examinations and reinforced in later life by the approval of peers, is very strong. Nevertheless, authority in science is authority with a difference. It rests much more on the ultimate appeal to personal experience and much less on any kind of organized threat system than does the authority of the state. The authority of science rests much less on the arts of persuasion, of ritual, of community reinforcement than does the authority of the church and religious organizations, although elements of this are not wholly absent, especially in the teaching laboratory, where the student is expected to get the experiment right, not to make a new discovery, and where the ritualistic aspects are quite strong.

The authority of science is based also on a remarkable ethic within the scientific community, which puts extremely high value on veracity and reserves its most extreme sanctions of exclusion for anyone caught violating this ethic and deliberately falsifying the result of an experiment. Geographical map-making is a good example of the extraordinary ethic and authority of science. Political map-making distorts frontiers. Religious map-making, like that of Dante, is rarely subject to the test of the surveyor. The faith that we have in atlases and topographical maps is so close to 100 percent that it never enters our minds that they could ever involve any deliberate deceit, although they might involve human failings and unwanted error.

Numbers have played a significant role in establishing the authority of science. They have done this in many ways. There are two broad modes of employment of numbers; through counting, and through measurement. In population statistics, demography, a good part of economic statistics, and such things as atomic numbers, we are essentially counting discrete elements. The number that we get is a whole number and it is not arbitrary. The other use of number is in measurement, in which what we are counting is an arbitrary unit, and therefore the meaning of the number is dependent on the definition of the unit of measurement. A stick may be 2 feet or 24 inches, or about 60 centimeters. All these numbers represent exactly the same length. This essential arbitrariness of the number used in measurement does not cause much trouble when it comes to the use of measurement in testing, as long as the unit is stable and clearly understood. It does cause trouble, however, as we shall see, when it comes to the translation of numbers into images of the "real world" of structures, topographies, and personal experience.

Number has been enormously useful in science for two basic reasons. One is that numbers are highly manipulable by the human mind. They can be added, subtracted, multiplied, and divided. They are the foundation for an important part of mathematics, though by no means of the whole of it. By means of manipulating numbers the human mind can handle n-dimensional structures which it is quite incapable of visualizing. While one can argue that numbers are a prosthetic device of the human mind rather than an expression of reality, such prosthetic devices are in no sense to be despised, and they have enormously increased the capacity of the human mind for developing images of highly complex realities beyond the images of personal experiences.

This would not have been possible had it not been for another virtue of numbers, that they are capable of mapping topological structures and relationships. A topographical map is a good example of this. Each point on it can be defined by two numbers, a latitude and a longitude. When we add a third number, the altitude, we have a perfect one-to-one relationship between these three numbers and the position of the point relative to the axis of the earth, some arbitrary point on the earth's surface, such as Greenwich, and its distance from the earth's center, or from the oblate pear-shaped spheroid known as sea level. Given the three coordinates of two points, then by our power of manipulating numbers, we can calculate the length of the line—or the arc—between them and its direction. Innumerable similar examples could be given.

The great power and usefulness of our capacity to manipulate numbers should not blind us to the fact, however, that numbers are much more a part of our internal map of the world than they are of the real universe outside us. The relation between maps and reality, as we have seen, is an enormously tricky problem in epistemology. At very crucial points we seem to come up against a virtually irreducible ignorance of what the real world is like. It may be, as James Jeans remarked, that God is a mathematician, that the real world, whatever it is, consists of numbers and abstract relationships and it is our topographical and topological images of it in our minds that are the illusions. This is a problem, however, that we can leave to the philosophers for all time to come.

The images of the world that we have in our minds are primarily images of shapes and structures, not images of numbers. I have speculated indeed—and this is no more than a speculation—that there are very few numbers in the real world. There is certainly e, π, the velocity of light and Planck's constant, valencies up to 4 (or maybe 6), and perhaps we should add atomic numbers up to about 132. As far as groups and clusters of things are concerned, we react to 0, 1, 2, 3, 4, perhaps up to $7+2$, and beyond, that numbers simply serve to map, often with a pretended accuracy far beyond either our needs or the realities, structures which are essentially topological rather than numerical in nature. It is often important to know whether A is bigger or smaller or the same size as B. It may be important to know whether it is much bigger or not much bigger, much smaller or not much smaller. There are rarely more than 5 to 7 grades (ordinal numbers) with most things that are significant in human life and experience. Quality is more important in human life than quantity, but quality is a very tricky concept. It is usually more than one-dimensional, but it can also be described with relatively few ordinal numbers. In human decisions usually all we need to know is three ordinal numbers: whether A is better, worse, or as good as B.

Physical scientists I am sure would be uneasy at the way this argument is drifting. Surely they will say the physical sciences are the exact sciences and exactness is derived from number, and indeed from measurement to a large number of decimal places. All measurement involves a probability distribution. When we say our stick is 24 inches, we mean that 24 inches is the mode of a probability distribution of its "true" length. Exactness is reducing the standard deviation of this probability distribution to levels that are psychologically satisfying, that is, the point at which doubt becomes psychologically equivalent to zero, whatever the number we give for its probability. The classical Michelson-Morley experiment is a case in point. If they could only have measured the orthogonal velocities of light with a standard error of 10 percent, the refutation of the hypothesis of ether drift would have been very much in doubt and Einstein would not have triumphed over Newton. Without exactness or sufficient exactness in measurement, crucial experiments may be impossible. Nevertheless the rage for measurement, especially in the social sciences, may have produced a spurious exactness. There is no point in getting an exact measure of something that is not really interesting or that is only loosely related to the thing we are really trying to investigate. This makes the difficulty of devising crucial experiments all the more acute.

Over against this position, however, there is the view that measurement can only be one-dimensional and that where the reality has many dimensions, as topographical and structural realities often tend to have, it tends to reduce a multidimensional reality to a single linear measure. Aggregative measures especially, may be extremely misleading and may result in much more spurious testing than the less exact, but more multidimensional attempt to test the essential structural similarities of n-dimensional structures. I cannot hope to resolve this argument. Certainly if the real world consists of n-dimensional topographical space, we are in real epistemological trouble with our essentially 3-dimensional minds. And we will constantly find that reality eludes us and that our imposed but illusory order is always breaking down into chaos.

The nature of ultimate reality, however, is perhaps less important to us than the workability of our current images of it in our minds. In some sense we are all pragmatists. We subject our images of the world to testing. If the tests succeed, our images are confirmed. This is why I have said that nothing fails like success because we do not learn anything from it. If our test fails, however, then we have to reorganize our image of the world, and the critical question is how do we reorganize it? If there is no alternative image, we rarely reorganize it, in spite of the failure of tests. A good example of this would be the impact of parapsychology. Suppose, for instance, that we could devise a large-scale test for psychokinesis, which would irrefutably prove its existence as a phenomenon. Theoretically, this should bring down most of the current images of the physical sciences in ruins. A phenomenon which does not obey the inverse square law violates one of the basic conceptual identities of the physical sciences. At the moment, however, there is no other model to go into, and we can be pretty sure that no matter how convincing the empirical evidence, the evidence will become compartmentalized and the physical sciences will emerge unscathed. It is only when there is an alternative image of the world that an experiment can be decisive. This was the case with the Michelson-Morley experiment.

The rest of this paper will be concerned with a much more mundane question than these large and difficult questions of the nature of the real world. This is the question of what might be called epistemological statistics, that is, to what extent does numerical information change people's minds, and change the character and quality of the images of the world which they possess, in regard to evaluations of certainty and of truth or error. This is a subject which has been seriously neglected by statisticians and by psychologists. I am prepared to argue indeed that epistemological statistics constitutes a whole missing discipline in science; that is, the study of the presentation of numerical and quantitative data in ways that favorably change people's images of the world in regard to workability and, one hopes, in regard to truth.

We see this problem even in the presentation of countable numbers. We have clear images of what small numbers mean. It has been argued indeed that we have probably genetically imposed skill of the recognition of clusters of objects up to plus or minus seven. We know what 1, 2, 3, 4, 5, 6 or 7 objects mean, and we visualize them easily. In terms of some rather nonrandom arrangements as in dominos, we visualize them easily up to 12 or more. We might have some difficulty in perceiving that a random group of 11 dots is different from a random group of 12. We certainly find it very hard to distinguish a random group with 87 dots from one with 88. Our capacity for number manipulation, however, and especially of multiplication, increases our capacity for visualizing numbers. We distinguish 64 from 63 very easily if they are arranged like a chess board. We can visualize a million as a chessboard of a thousand squares a side and a billion as a cube a thousand inches to a side. Then 999,999,999 would be that cube with a small bite out of it! A million people could stand in a square a thousand feet on a side. The population of the United States could stand in a square about 15,000 feet or about 3 miles to a side. The whole population of the world could stand in a square about 13 miles to a side, far less than a county. And all human beings who have ever lived could stand in a square about 50 miles to a side or even more dramatically, could be encased in coffins in a cube about a mile and a half to a side.

Large numbers can be dealt with by division as well as by multiplication. This is why per capita figures are so useful. The gross national product (GNP) of the United States or of the world is difficult to visualize in terms of trillions of dollars, but a per capita GNP of $6,000 or of $1,000 or of $200 is very easy to visualize. Proportions, likewise, expressed in terms of percentages, are relatively easy to visualize, though we have to be careful when visualizing proportions and percentages that we do not lose sight of the total number. Percentages can be very misleading if we do not know what the base is. Russian statistics have been notorious for this. There is the famous story indeed of the Russian village in which 50 percent of the men married 10 percent of the women, there being 2 men and 10 women. Many structures, however, can be visualized more easily in terms of proportions or percentages than they can in terms of absolute figures and the judicious combination of the absolute and the relative is a problem in statistical art for which no simple formula can be given. The use of logarithms in this connection has considerable potential of a fairly sophisticated kind.

In visualizing the meaning of measurement numbers, it is extremely important that the unit of measurement correspond to something that is easily visualized in ordinary human experience. Traditional measures, indeed, tend to be such even though they often become obsolete with the changes in common experience itself. Thus, the "foot" is clearly related to the length of the human foot. The yard is supposed to be the distance from the tip of the nose to the end of the outstretched arm used in measuring cloth. The inch is the thickness of the average thumb. The hand, sometimes used in measuring horses, is of course the average distance across the hand. The bushel is a basket, the pint is a jug and the jill is a cup, the furlong is a "furrow long," the acre is what could be plowed in a day. Even as the rod, pole, or perch is about the longest convenient rod a surveyor could carry and a chain about the heaviest chain he could carry.

The metric system was careless in assigning units that had human reference. The meter is close enough to the yard to have human significance, although 2 yards is a much commoner human height than 2 meters. The meter is perhaps a little long for a pace, the average of which may be close to a yard, though the kilometer is certainly close to a thousand single paces (of a 2-meter human), and the mile is a thousand double paces of a shortish Roman soldier. The centimeter perhaps could be defended as the width of the little finger, the decimeter perhaps as a very small hand. The gram is barely a thimble full and is far too small to be a psychologically satisfactory basic unit. The dyne and the erg are preposterously small by human standards, almost below the threshold of perception. On the other hand, the British thermal unit (BTU) is hard to defend also as it is an incredibly rare human experience to heat a pound of water from 59.5 degrees to 60.5 degrees Fahrenheit. Temperature is tricky; freezing is not a good zero psychologically because we experience temperatures below this. Fahrenheit himself evidently has set his zero at the lowest temperature that he could achieve with ice and salt. On the other hand, there is something psychologically satisfying about the Fahrenheit zero which is damned cold and anything colder than that, below zero, is very cold indeed. The Fahrenheit unit was designed, I think, to make blood heat approximately 100, but did not quite get it right. The real psychological difficulty here is that our perception of temperature differences varies with the temperature itself. We are quite conscious of the difference in room temperature between 68 and 70. We would not be so conscious between 0 and 2 degrees or 100 and 102 ambient temperature. This seems like a rather insoluble problem, but it is fortunately one that we probably do not have to solve.

It is clear of course that all measurement if it is to be uniform has to be arbitrary. The critical question is whether the arbitrary unit that we select is close enough to some common human experience to make it easily visualized. The size of the unit and the multiples of it also are of some importance. A unit should certainly not be so small as to be below the just noticeable difference in perception, nor so large that we have an urge to subdivide it. Custom and constant use, of course, can give meaning to almost any unit, but there is still a real problem as to what unit causes the least trouble in visualization. These units may need to change also as common human experience changes. I have argued, for instance, in the case of energy that neither the metric nor the British system are very satisfactory and that perhaps we should use a unit like the "bulb-hour," a bulb being a 100-watt bulb, of which we all have experience. If we are told that the rate of per capita energy use in the United States is 113 "bulbs," we can visualize ourselves with these glaring around us and this gives us a much more vivid idea as to what the number means than when it is expressed in ergs or in BTU's, still more in "quads," which is a quadrillion BTU's. There is need for serious psychological research here, which as far as I know has never been done, into how people relate their ordinary day-to-day experience to the units of measurement which they use. Research of this kind ought to come up with at least a reasonable range of appropriate units in all fields, a range, of course, which might well be a function of the total culture in which people live and so might vary from time to time and from place to place.

An even more controversial question than the selection of units is the question of the counting system itself. Outside of the binary system, which has a certain austere necessity about it and is, of course, the basic system of computers, the counting system is quite arbitrary and in the history of the human race a number of different counting systems have emerged, many of which survive to this day. The decimal system is supposed to have developed because we had ten digits on our two hands, though this does not seem to be a wholly satisfactory reason, for, after all, we have two ears to add to this if we wanted to make it duodecimal. The binary system and its extensions into a scale of 4, 8 or even 16 never seems to have been extensive, although there are traces of it, of course, in the duodecimal system and traces of it also in subdivision. The subdivision of the inch into halves, quarters, eighths, sixteenths and thirty-seconds has some real psychological foundation behind it as it seems much easier for the eye to accommodate this division than the division, for instance, into tenths. Counting by sixes, by twelves and by sixties is very old and certainly goes back to Babylonia and remains today particularly in a measurement of time, although the French metricists did attempt, fortunately unsuccessfully, to decimalize the clock (see Chapter 5). The 24 hours a day, or 12 hours for the half day, reflect the ancient duodecimal system of counting. Vestiges of this also remain in the 12-inch foot and, of course, in the now obsolete 12-penny British shilling. A tendency to combine the decimal and the duodecimal is apparent; the 60-minute hour and the 60-second minute would reflect this as did the old British pound with its 20 shillings.

The main advantage of these non-decimal systems is in division. The 24 hours in a day can easily be divided into half—12 hours; and a third—8 hours, which represents an ancient division into sleep, work and leisure. It can also be divided into quarters, eighths and so on. The 60-minute hour divides into halves, quarters, thirds, fourths, fifths, sixths, tenths and twelfths. The finer divisions may not be important, but division into thirds and into quarters is extremely convenient. A decimal clock of 10 hours would have no quarters, no thirds. Anyone who has traveled with two companions knows how every accounting comes out to a third of a cent.

What would be the most convenient number system is surely not an absurd subject for psychological research. The binary system has a serious disadvantage in that it takes about three times as many symbols as the decimal system. The duodecimal system would require the invention of two new names for the numbers presently called ten and eleven, and this might be awkward. An octimal system based on 8 would have a good many advantages. A seximal system based on 6 might be better as it would combine the binary and the ternary. Whether the advantages would be large enough to make a change worthwhile is, of course, open to question. The same question could be asked of any change in metric including the change from the traditional to the "metric system."

It may be argued that all this is a tempest in a teacup and that it really matters very little what kind of metric or system of measurement we use. The inconveniences of the bad ones are so small that it is not worth making a fuss about. This is an argument to be taken seriously. The only case for the French metric system is that it has become sufficiently universal so that there are real advantages in making it completely universal. It cannot claim the slightest scientific validity as its units are not based on any natural units and are psychologically not even particularly convenient. The decimal system is one of the less convenient systems of counting, though by no means the worst. The only argument for it is that when it doesn't really matter what we do, it is convenient to have everybody do the same thing, so let's all join the party. I have joined Professor David Bartlett, however, in a plea for a second look, simply from the point of view of science.

There is also the argument that there is pleasure in cultural diversity. It should be perfectly possible for each society to use the French metric system in its international relations and some traditional system locally. The importance of traditional metrics in the metaphors and usage of local languages is another argument for them. If the speakers of English, for instance, forget what it means to inch forward, or to eat a peck of dirt, or to hide a light under a bushel, or to go the second mile, or to be buried in God's acre, some richness would have gone out of the language and the life. The traditional measures may have a symbolic value not to be given up lightly. In Colorado, for instance, there is a magic about the mile-high city that 1.6 kilometers would not have and a similar magic about the 14,000-foot peak that no number of meters could emulate. There is no reason indeed why a traditional metric and a more universal metric could not exist side by side and be used for different purposes. The capacity of the human brain could certainly encompass this degree of richness. If we insist on a universal metric, however, we do need to examine very carefully its validity both in terms of natural units and in terms of the translation of numbers into psychological images.

The latter problem becomes more acute as we move into more complex systems. It is unlikely that any great decision or any catastrophic mistake will rest upon whether we measure length in feet or in meters. All that is at stake here is a certain degree of convenience. Misapprehensions, however, about the basic structure of human potentials of behavior and social interaction may be quite catastrophic. Here the proper learning of meaningful numbers may be of great importance in human decision, especially where this can affect our whole taxonomy and our evaluation structures. A good example would be the concept of race. Once we know that racial differences represent an extremely small fraction of the human genetic information structure and that the significant genetic differences within races are greater than those between them, our evaluation of racial differences would certainly change. Similarly, ignorance of the dimensions of the economy can profoundly affect our evaluation of different forms of it. I once asked a group of worthy and well-educated citizens each to estimate the proportion of the national income of the United States that goes to labor. I received answers ranging from 10 percent to 90 percent in an almost uniform distribution. This is a number, however, which with a small error is in the public domain. It is of the order of 75 percent; yet this "public fact" was completely unknown to most of these people. Clearly our political views would be affected profoundly if we think that 10 percent of the national income goes to labor or if we think that 90 percent does. A survey not too long ago revealed that a large proportion of people in this country believe we have about 500 to 1000 years' supply of oil and gas. It is not surprising there is no sense of an energy crisis in the general public. The question of what numbers are epistemologically significant (that is, change people's minds) is, therefore, very far from being merely academic; it has high practical consequences. The debate about the

virtues of the French metric system may have only minor significance for the human race. The larger question of the meaning,
learning, and appreciation of significant numbers is by no means
trivial and it may make all the difference between survival and
non-survival.

4.
Natural Units:
An Alternative to SI

David F. Bartlett
*Associate Professor of Physics, University of Colorado,
Boulder, Colorado*

"During the last 100 years, there have been many proposals for basing measuring units on physical constants of nature, such as the speed of light, the gravitational constant, and the atomic constants. Albert Einstein once proposed that the diameter and mass of the hydrogen atom and the speed of light be the primary units of measurement, from which all other units could be derived. None of these proposals has ever been adopted by any nation.

"Some have argued that, no matter what base units are used, their multiples and submultiples be related by binary numbers: the powers of 2 and the fractions 1/2, 1/4, 1/8, 1/16, and so on.

"In principle, almost any precisely defined and consistent measurement system could serve us satisfactorily. In practice, however, it is unrealistic to consider for general use any choice of measurement system that is alien to our culture or to that of the rest of the world. The U.S. therefore really has only two practical alternatives: either to allow its measurement system, which includes some metric units, to develop without overall design, or to elect as a society to adopt the measurement system that has virtually achieved worldwide acceptance and to work out a policy and program for changing to it."
    —*A Metric America* (U.S. National Bureau of Standards Spec. Pub. 345 [1971])

Thus did the report of the National Bureau of Standards introduce "natural units" to the Congress. In so quickly disposing of this solution to the units problem, however, the report may have slighted a genuine possibility. Suppose our units of measurement continue to develop "without overall design." The end point of even this evolutionary process has always been assumed to be eventual—and permanent—acceptance of the International System of Units (SI). This system, however, has not yet been fully accepted, even by European scientists and engineers. Conversely, the arguments for a natural system are becoming more compelling as technology exploits more fully the relations inherent in the atomic structure of matter. It is conceivable that in the generation it would take to convert without plan a worldwide consensus favoring a natural measurement system will emerge. Unfortunately, this possibility would be postponed indefinitely if the United States converts purposefully in the near future.

Believing that this option for a more optimal system of measurement should be kept open, 22 faculty members of the Department of Physics and Astrophysics at the University of Colorado signed a petition in 1975 urging rejection of the then pending metric conversion legislation (see appendix II to this chapter). People signed for various reasons. The following gives one person's reason for being against metric conversion.

Problems of the SI System

Plastered across the walls of many schools and laboratories is an NBS poster entitled "The Modernized Metric System (The International System of Units—SI)." From a distance, the poster encourages the novice to believe that if he can master the 7 base units (meter, kilogram, second, ampere, kelvin, candela, and mole) he will be well advanced in understanding at least the vocabulary of measurement. Alas, the fine print dispels this illusion. Here he finds derived units such as the volt, joule, and pascal, all with coined names. "If these units are so readily derivable from the base units, why do we need special names?" he may ask. Then he notices the table of 16 prefixes ranging from atto$= 10^{-18} = 0.000$ 000 000 000 000 001 past milli, centi, and kilo all the way to exa$= 10^{18} = 1$ 000 000 000 000 000 000. "If this system is so practical, why do we need so many prefixes?" Fortunately, the poster tells where more information can be had, so he sends for the definitive booklet, *The International System of Units* (NBS Spec. Pub. 330; see appendix to this book).

This informative work gives most of the common physical quantities and their SI units. For example, a table "SI Derived Units with Special Names," has 18 entries ranging from hertz for frequency to gray for absorbed dose. A critical look at this table shows that it is not merely laziness that has led to these special names. Rather the *base* units *alone* are simply not helpful. We may visualize a volt as a watt per ampere or maybe as an ampere ohm, but never as a meter-squared kilogram per second-cubed ampere! (For the relationship between the base units and other derived units see Figures 1a-1d.)

Altogether, the NBS booklet lists 47 distinct derived quantities. Of these, only 15 are conveniently expressible directly in terms of the base units. These 15 are admittedly the most common (such as volume, which can be readily related to a cubic meter). But ironically the 15 are so common that they are generally visualized as entities in themselves quite apart from the base units. Just as we might visualize a gallon as a jug of cider rather than 231 cubic inches, so the Frenchman would think of a liter as a bottle of wine rather than a thousandth of a cubic meter.

That the base units do not provide a generally useable key to the derived ones is candidly admitted in the introduction to NBS-330:

> SI units are divided into 3 classes: base units, derived
> units, and the supplementary units. From the scientific.
> point of view division of SI units into these three classes
> is to a certain extent arbitrary, because it is not essential
> to the physics of the subject.

Granted that this statement is not comforting to the scientist or teacher, can it at least be said that the resulting system is practical? Do the SI units have magnitudes which permit one to speak of a few this or a few that? Yes and no. The system is by design

1a. BASE UNITS

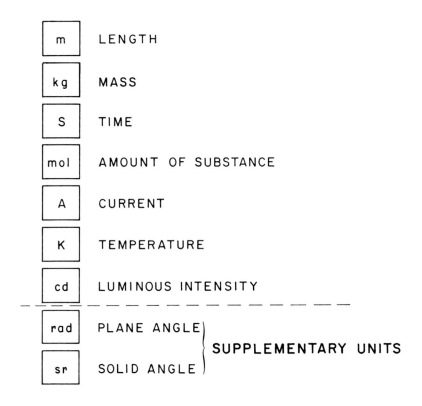

m	LENGTH
kg	MASS
s	TIME
mol	AMOUNT OF SUBSTANCE
A	CURRENT
K	TEMPERATURE
cd	LUMINOUS INTENSITY
rad	PLANE ANGLE
sr	SOLID ANGLE

SUPPLEMENTARY UNITS

Figures 1a-d. SI Units: These figures show the relationship between the base units and the derived ones listed in NBS-330. The presentation is suggested by a design of H. J. Milton. (In particular Figure 1c "Derived Units with Special Names," is essentially a copy of Milton's design.) The connection between the base units and the "Directly Derived" units on Figure 1b is very clear. The links on Figure 1c and on Figure 1d, "Derived Units Expressed by Special Names" are more tortured. I call units "practical" if they are generally used without prefixes.

1b. DIRECTLY DERIVED UNITS

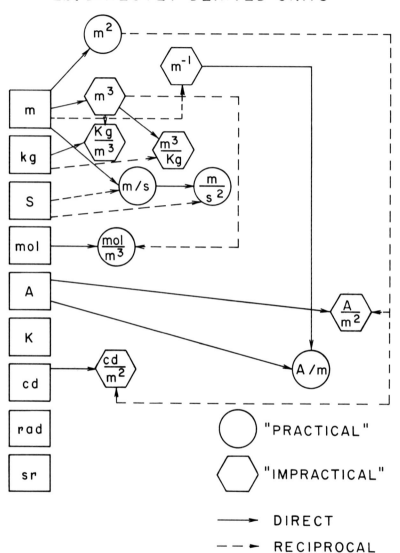

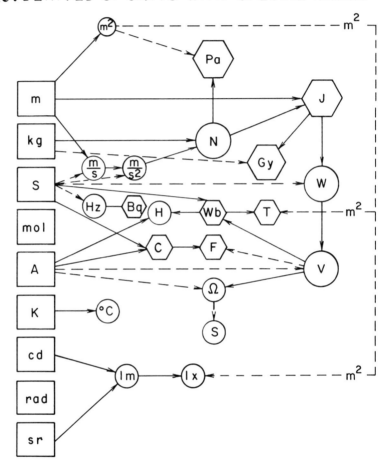

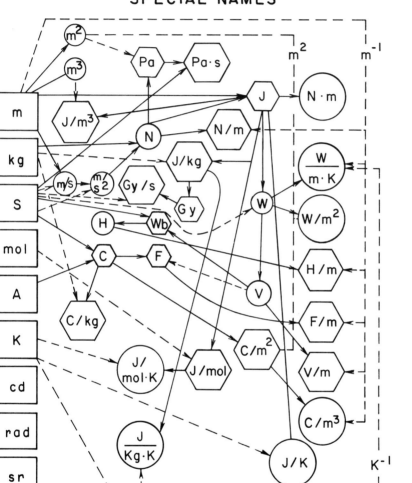

especially well-suited to the needs of the electrical engineer. The volt, ampere, and ohm are admirably practical units. The unit of capacitance, the farad, is not. The millionth of a farad (microfarad) is effectively the practical unit. The SI unit for magnetic field, the tesla, at 10,000 times the earth's field, is so large that it is nearly always replaced by the non-SI unit, the gauss.

The practicality of the SI units is questionable in other fields as well. One gasps for breath at an atmospheric pressure of *only* 50,000 pascals; yet just a *few* grays (the new SI unit for absorbed radiation dose) is sufficient to make one very gray indeed. Overall, the SI units are practical for only about half of the 47 common quantities listed in NBS-330.

To make the system workable we must use 16 prefixes. These are decimal multipliers such as centi = 0.01. Even this large number is insufficient, a fact which the International Congress of Weights and Measures (ICPM) reluctantly admits by adding a few more each decade. One could live with these prefixes if all one had to do was to specify quantities (e.g., the wheat production in the United States was 60 teragrams last year). But most of us have to make computations (e.g., to figure the grain production per unit area). We soon find ourselves making conversions (e.g., tera = 10^{12}). The chance for error in these conversions is large, a fact which is acknowledged by the International Standards Organization in their Recommendation R1000-1969 (E):

> In order to avoid errors in calculations it is essential to use coherent units. Therefore, it is strongly recommended that in calculations only the SI units themselves be used and not their decimal multiples and submultiples.

Finally, the range of powers encountered in acoustics or electrical amplification is so great that it has been found convenient to use a logarithmic unit, the decibel (multiplication by 10 is equivalent to the addition of 10 db). The ICPM is ill at ease with this measure (since 10 db + 10 db = 13 db instead of 20), and observes tersely that the db is not an SI unit. Many feel, however, that the compact notation afforded by the decibel should be extended to quantities other than power rather than be curtailed.

To recapitulate, the SI system is deficient because even with its large number of base units (scientists are used to 3, not 7) it fails to provide a ready means for either visualizing the derived units or for assaying their practicality. In addition, the mixture of prefixes, scientific notation, and standard arithmetic notation leads to confusion in calculations.

That this system needs improving is acknowledged everywhere. Even the Metric Conversion Act states:

> The metric system of measurement means the International System of Units . . . interpreted or modified for the United States by the Secretary of Commerce.

Will the few peripheral modifications which NBS is presently considering be satisfactory to provide a measurement system for the indefinite future? Personally I doubt it. To find a truly satisfactory system, I believe we should look carefully into the longstanding proposal for basing our units on the physical constants of nature.

Natural Units

> In forming the existing artificial [metric] series of systematic units, it has been usual to regard the units of length, time, and mass as fundamental and the rest as derived; but there is nothing to prevent our regarding any three independent members of the series as fundamental and deriving the rest from them. Nature presents us with three such units; if we take these as our fundamental units instead of choosing them arbitrarily, we shall bring out quantitative expressions into a more convenient, and doubtless a more intimate, relation with Nature as it actually exists.
> —G. J. Stoney in *Philosophical Magazine* (1881)

In referring to the metric system as artificial, Stoney meant no slur. He merely noticed that by quantifying nature with artifacts suited to human dimensions and perceptions we make that description unduly cumbersome. Rather we should let nature herself suggest the appropriate tools. We might hope that these constants would take the form of a fundamental mass, a fundamental length, and a fundamental time, by which all physical phenomena could be readily described.

The theory of relativity, however, has shown that none of these concepts is really appropriate. Mass is not strictly additive; when two chemicals combine, some matter is lost as energy. Similarly, length and time intervals depend on the relative motion of source and observer. Nature, fortunately, has provided us with appropriate variables. Preeminent among these are the velocity of light (c) and Planck's constant (h). (Planck's constant relates the energy of a quantum of light to its frequency.) By setting these constants equal to one, we can achieve Stoney's dream of a system of measurement brought into "more intimate relation with nature as it actually exists."

Theoretical physicists have long realized this simplification to be helpful, but their enthusiasm has yet to spread to other scientists, engineers, and the public generally. Natural units have remained within the preserve of physicists for three reasons.

First, c and h are only two constants. To complete Stoney's dream, a third constant is needed. Stoney and others have offered the universal constant of gravitation, G, as this third constant. Such a suggestion is, however, inappropriate for a generally useful system of units. This is because most measurable phenomena are ultimately related to atomic processes. Such processes are now well understood, but their description does not involve gravitation. In a way it is fortunate that G is not better qualified, for by setting h, c, and G all equal to one, the units for mass, length, and time are all uniquely determined. Now (as metric conversions elsewhere indicate) man will accept a change in his units for length and mass. For earthbound people, however, the natural unit of time is and always has been the mean solar day (although for precise work even the unit for time is now defined by an atomic transition).

Second, until recently the constants c and h could not be related to the kilogram, meter, and second precisely enough to permit their replacing these units in conventional measurement. A century ago, the speed of light c could be measured only to an accuracy of about 3%. Thus, if Stoney wanted to define a unit distance as the distance light goes in a unit time, then the unit distance would only be determined to an accuracy of 3%. Clearly no one would be happy with such a poorly defined ruler. Now, however, the speed of light has been measured so well that the meter itself may soon be defined by how far light travels in a given time. Indeed, scientists at the National Bureau of Standards have recently suggested that *all* the base units of the SI system may soon be determined from the atomic standard for time via defined values for such atomic constants as c and h. Of course, NBS is thinking of a definition such as c = 299,792,458 meters per second. The natural system offers the simpler definition: c = 1.

The third problem is one of nomenclature. The velocity of light is so large that the velocity of ordinary objects seems very small by comparison. What motorist would exchange 60 mph for 90 billionths (of the velocity of light)?

The problem of nomenclature can be solved by the development of an automatic way of representing repetitive multiplications or divisions by 10. For instance, the product of 1 multiplied by 10 four times may be written in the scientific notation (now being taught in our junior high schools) as 10^4. Similarly, a billionth, which is the quotient of 1 divided by 10 nine times may be written as 10^{-9}. But "ten-to-the-minus-nine" is even more awkward than a "billionth." Ironically, computers are slowly humanizing scientific notation. The exponential notation of computers is so widely used that it is the notation of choice in the list of conversion factors for the Standard Metric Practice Guide of the American National Standards Institute. In this notation, ten thousand is written as E + 4 and a billionth as E-9. It takes little imagination to visualize the simplest exponential notation where 10,000 becomes Positive 4 or P4 for short and 1/1,000,000,000 becomes Negative 9 or N9.

For the motorist, N9 would effectively become the "bench-mark" to which his car's speed is referred. He would no more need to visualize N9 as a billionth of the speed of light than today's motorist needs to visualize a mile as *milia passuum* or a thousand paces. In time, "N9" could be omitted from highway signs entirely just as "mph" is generally omitted now.

A simplified exponential notation would make natural units acceptable in our daily life. The natural structure of these units would, I believe, make the basis of our technological society more accessible to the general citizen than it is at present. Let me be specific in indicating how by setting $c = h = 1$ we can overcome the deficiencies of the SI system.

It may appear surprising that one unit, time, could provide a more generally useable aid to understanding physical phe-nomenon than can the 7 base units of the SI system. The secret is in making the "natural" choice of setting $c = 1$ and setting $h = 1$. The former choice permits one to measure lengths in terms of how far light travels in a day or the "light-day." The latter permits energies to be measured in the reciprocal of the same unit. Then Einstein's equation, $E = mc^2$, shows that mass as well as energy can be measured in terms of reciprocal days or day^{-1}. Having deter-mined the dimensionality of mass, length and time in terms of the day, all other physical quantities can be directly related to time. For instance, the unit of mass/volume or density would be $day^{-1}/day^3 = day^{-4}$.

Such a scheme would be of purely formal interest were it not for the fact that by far the majority of physical processes are directly related to atomic phenomena. We now know that such diverse things as the voltage of a car battery, the density of a solid, or the wavelength or visible light are all determined by atomic forces. The scale for these phenomena is set by only two param-eters; namely, the charge and the mass of the electron. In natural units, the charge of the electron is only a little less than one, so that effectively the scale for atomic processes is set by the mass of the electron which equals 10^{25} or P25 day^{-1}.

Because of the atomic nature of physical processes, most quan-tities which have a dimensionality of day have a measure of about

N25. Those which have a dimensionality of day^2 have a measure of N50, etc. Thus the wavelength of visible light is about 2 N20 day; the voltage of a car battery is 30 P20 day^{-1}; the density of water is 2 P98 day^{-4}. This correlation between dimensionality and practical range is so strong that one can safely drop the unit "day" from the expression and, for instance, quote voltages simply as so many P20. When the range of a quantity is large it may be convenient to use a completely logarithmic notation and write P20.3 instead of 2 P20. In this manner decibels are naturally incorporated into the system.

A system which is ideal for atoms but not for humans would hardly be practical. Fortunately, the light-day is close to 10^{15} thumbreadths or to 10^{14} times the length of the naked human foot. Therefore the natural units for length, the N15 and N14 are suitable replacements for our present inch and foot respectively. (See Figures 2a-c.)

Such a system, I believe, overcomes the defects of the SI system. No longer is one awash in a sea of prefixes and the names of dead physicists. A natural system combined with a simple exponential notation permits us to fulfill the dream of the programmer Lafay, "Let us count with numbers, not names."

Binary Units

In talking to others about the metric system, I have found more interest in its mathematics than its physics. Many are concerned that by going metric we will bind mankind even more strongly to the decimal system than we are now. The superiority of the binary system to the decimal one is generally conceded in three separate fields.

First there is the progression of sizes. Our customary progression of cup, pint, quart, half gallon, and gallon reflects the commercial preference for packages spaced apart by a factor of 2.

In some cases, a factor of 2 may be too small and 4, 8, or 16 may be used. In others, such as the gauge of sheet metal or the diameter of wire, steps as small as the square or the fourth root of 2 naturally suggest themselves.

RANGE OF
EVERYDAY QUANTITIES

	N60	N40	N20	P0	P20	P40	P60
SPEED							
TIME					TEMPERATURE		
LENGTH						MASS	
AREA							
VOLUME							

Figure 2a. Natural Units: Everyday quantities as displayed in a natural system of units having h = (Planck's constant) = c (velocity of light) = 1. The unit of time is taken to be the day. Everyday quantities would include

Speed:	*"ninen"*	*N9 ≈ 1 ft/sec*
Length:	*"fifteenen"*	*N15 ≈ 1 inch*
Area:	*"thirtyen"*	*N30 ≈ 1 sq inch*
Volume:	*"forty-fiven"*	*N45 ≈ 1 cubic inch*
Temperature:	*"fip"*	*P15 ≈ 1°F*
Mass:	*"fourp"*	*P54 ≈ 3 oz avdp.*

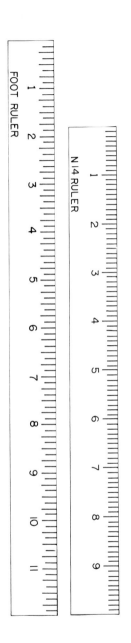

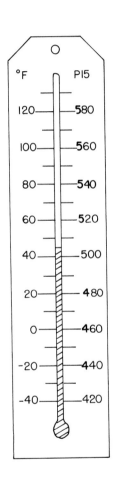

Figure 2b. A N14 ruler compared to a foot ruler, and a thermometer marked in degrees Fahrenheit and P15. Used with the permission of the American Journal of Physics.

Second, carpenters have long used a system where the inch is divided in a binary fashion. The advantage of such division was aptly expressed by John Quincy Adams in his 1821 *Report to Congress on Weights and Measures*:

> A glance of the eye is sufficient to divide material substances into successive halves, fourths, eighths, and sixteenths . . . But divisions of the fifth and tenth part are among the most difficult than can be performed without the aid of calculations."

Finally, today's computers are intrinsically binary machines. They can, of course, be programmed to accept decimal numbers at the expense of more input hardware or to calculate in decimal at the expense of more memory locations. Neither expense would be entailed if our number system were binary rather than decimal.

Our number system is decimally-based, however. I am prepared as most are to accept this as an accomplished fact. That is, I was prepared to accept the permanence of the decimal system until reading an article by R. O. Whitaker, "Is the Arabic-Decimal System Adequate to Meet Today's Needs?" In this interesting article, Whitaker discusses how one can assemble a base-16 number out of four binary strokes which can either be made or not. With a slight modification this "assembler digit" would be directly compatible with the decimal light-emitting diode displays now so common on miniature calculators.

Experience in other countries has evidentally indicated that the costs of metric conversion are not nearly as great as initially feared. Would not the long-term benefits of a binary system justify making a study of the costs and benefits of a slow transition to that number system?

Commercial Problems

The reader who has followed so far may well be wondering what all this has to do with the price of wheat in Chicago. Isn't our real problem how to join the rest of the metric world and to streamline our industrial practices in the bargain?

The hooker is the little word "and." It appears that accepting foreign practices is often in conflict with perfecting our own. Take the pathfinding experience of the Optimum Metric Fastener System (see Chapter 2). In 1971 this program was boldly announced as an attempt to find the best screw and nut design. The product was to be described in metric terms but was not to be predetermined by an old technology, either metric or ours. Indeed, a rational size progression suggested 25 sizes of which 9 should be completely new.

The European reaction to this proposal was to say that the United States could develop whatever fasteners it pleased—so long as they mated with the existing ISO metric coarse series. Such a declaration ensured the indefinite continuation of a non-optimum progression of sizes.

The Optimum Metric Fastener System excelled elsewhere, in gaging practice and head design. One can only wonder, however, how much better the final product might have been if European and United States manufacturers were confronted with a third measurement system, one which was neither metric nor English. Then all countries could cooperate on developing optimal industrial modules without offense to national pride or prejudice.

We have been told by some that the International System of Units is the best that can be devised; it isn't. We have been told that the decimal system is unchangeable, but the arguments for changing it appear to grow stronger with time rather than weaker. Finally, we have been told that commerce needs the change, but perhaps a genuinely new system would be better.

It appears to me that this is not the time for the United States to accelerate its conversion to the International System of Units.

Bibliography

Specific References in Paper

Adams, John Quincy. 1870. Report to Congress on weights and measures (1821). Reprinted in *Metric system*, ed. C. Davies. New York: A.S. Barnes.

Barbrow, L.E. 1976. *What about metric?* U.S. National Bureau of Standards Consumer Information Series No. 7.

Bartlett, D.F. and Zafiratos, C.Z. 1976. Metrication and motherhood. *Physics Today* 29, no. 2: 15.

Belford, R.B. 1975. International overview of metric fasteners. Paper presented at Metric Conference of the American National Metric Council, 18 March 1975, Washington, D.C.

De Simone, D.V., ed. 1971. *A metric America.* U.S. National Bureau of Standards Special Publication 345.

Hellwig, H., Evenson, K.M., and Wineland, D.J. 1978. Time, frequency and physical measurement. *Physics Today* 31, no. 12: 23-30.

International Standards Organization R-1000-1969(E) recommendation 3.1.

Mallen, S.E. 1974. The optimum metric fastener system. *Automotive Engineering* 82: 21. See also chapter 2 of this volume.

Page, C.H. and Vigoureux, P., eds. 1977. *The International System of Units (SI).* U.S. National Bureau of Standards Special Publication 330. See also Appendix to this volume.

Stoney, G.J. 1881. *Philosophical Magazine* 11: 381.

General

Ambler, E. 1971. *SI units, the philosophical basis for the base units.* U.S. National Bureau of Standards Technical Bulletin 55.

Bartlett, D.F. 1974. Natural units, are they for everyone? *American Journal of Physics* 42: 148.

Carrigan, R.A. 1978. Decimal time. *American Scientist* 66: 305. See also chapter 5 of this volume.

Cohen, E.R. and Taylor, B.N. 1974. Specification of the physical world. *Dimensions* 58, no. 1: 3-6.

Danloux-Dumesnils, M. 1969. *The metric system: a critical study of its principles and practice.* London: Athlone Press.

Appendix I to Chapter 4

Range of Physical Quantities

In the text I stated that in a natural system of units a typical value for a given physical quantity tended to be inversely correlated with the temporal dimension of that quantity. Loosely speaking, this relationship results from the fact that the time scale for atomic processes is very short when compared to the day. The figures that follow (3a-f) support this correlation in a general, but not detailed fashion. To appreciate the range of a physical quantity it is clearly necessary to know more than the mass of the electron!

Surprisingly, not too much more information need be given. In addition to the basic definitions of a natural system of units, namely:

> 1 = velocity of light in free space
>   = Planck's constant
>   = Boltzmann's constant
>   = permeability of free space
>   = permittivity of free space
>   = mean solar day,

we need but a few experimentally determined quantities. Knowing the mass of the electron m *and* its *charge* e we can estimate the magnetic moment of the electron, μ_B, the radius of the hydrogen atom a_o, and R_∞ the energy required to ionize hydrogen (or separate its electron from its proton).

Atoms heavier than hydrogen are much more complicated to analyze. Fortunately, the competing forces in heavy atoms—attraction of the electrons to the nucleus and repulsion from each other—are so balanced that we can make the crude approximation that for *any* atom its radius is the Bohr radius a_o and its first ionization potential is the Rydberg R_∞. Thus the Rydberg determines the scale for chemical energy exchanges; the Bohr radius determines the spacing of atoms in a solid.

The nucleus of an atom is much more massive than an electron. To estimate the density of matter we shall need the experimentally determined mass of a typical nucleus, say the mass of the carbon nucleus, m_C. We shall also need m_C because the vibrational energy between atoms in a molecule or nearest neighbors in a solid approximates $(m/m_C)^{1/2} R_\infty$.

Finally, to connect atomic phenomena with human observations it is often necessary to use samples of many atoms. (What

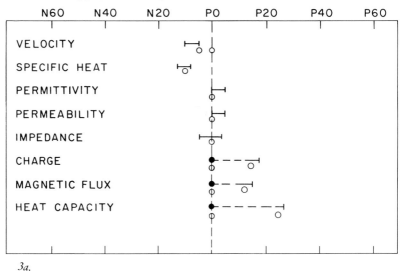

3a.

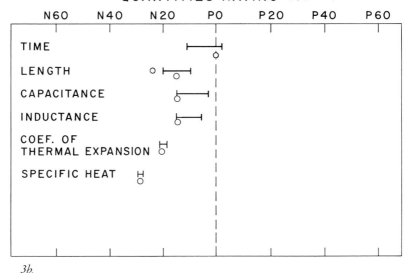

3b.

Figures 3a-3f show the range of Physical Quantities displayed according to their dimensionality in time. Note strong correlation between dimension of quantity and observed range. Open circles indicate a "natural atomic construct"; e.g., for mass: mass of carbon nucleus m_c = 2 P29.

RANGE OF
QUANTITIES HAVING <T>=-I

N60	N40	N20	PO	P20	P40	P60

ACCELERATION

FREQUENCY

TEMPERATURE

VOLTAGE

CURRENT

ENERGY

MASS

3c.

QUANTITIES HAVING
<T>=2   <T>=-2

N60	N40	N20	PO	P20	P40	P60

AREA   ELECTRIC FIELD   (E)

ELECTRIC FLUX DENSITY   (D)

MAGNETIC FIELD   (H)

MAGNETIC FLUX DENSITY   (B)

THERMAL CONDUCTIVITY

POWER

FORCE

3d.

QUANTITIES HAVING

⟨T⟩=3   ⟨T⟩=-3

N 80	N 60	N 40	N 20	P 0	P 20	P 40	P 60	P 80

VOLUME○

CURRENT DENSITY

CHARGE DENSITY

CONCENTRATION

SURFACE TENSION

DYNAMIC VISCOSITY

3e.

QUANTITIES HAVING

⟨T⟩=4   ⟨T⟩=-4

N100	N 80	N 60	N 40	N 20	P 0	P 20	P 40	P 60	P 80	P100

IRRADIANCE

PRESSURE

SPECIFIC VOLUME   DENSITY

3f.

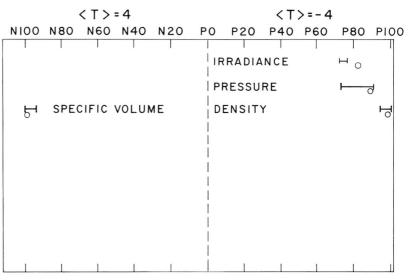

motorist would be satisfied with just a few molecules of C_8H_{18}?) This connection is facilitated by Avogadro's constant which specifies the number of atoms in 12 grams of carbon.

To obtain a crude understanding, then, for the range of physical phenomena we shall need the three experimentally determined quantities:

$$m = m_e \left(9 \times 10^{-31} \text{ kg}\right) = P25.0$$
$$m_C \left(2 \times 10^{-26} \text{ kg}\right) = P29.4$$
$$e \left(1.6 \times 10^{-19} \text{ C}\right) = (2/137)^{1/2} = N0.9$$

together with the derived quantities:

$$\mu_B \left(9 \times 10^{-24} \text{ J/T}\right) = e/4\pi m = N27.0$$
$$R_\infty \left(1.1 \times 10^{7} \text{ m}^{-1}\right) = me^4/8 = P20.5$$
$$N_A \left(0.012 \text{ kg/m}_C\right) = P23.8$$
$$d_o = 2a_o \left(1.1 \times 10^{-10}\text{m}\right) = N23.4$$

The above parameters are all that we shall generally need to form a representative value or "natural construct," for any physical quantity. For some special situations it will be helpful to use such terrestrial quantities as average atmospheric temperature and pressure and average acceleration of gravity.

The following tables give a list of various physical quantities arranged according to their dimensionality in time. Also indicated is the appropriate SI unit and its equivalent natural value as well as the range of the physical quantity. The extremes indicated are not candidates for the Guinness Book of World Records. They are rather extremes of general laboratory practice.

I beg the scientists' indulgence for the crudeness of the natural constructs. They are not intended to give a thorough indication of the complexity of real matter. It is hoped, however, that these constructs will take the reader one step beyond the SI system. In this system the base units alone are often not an aid in visualizing or in determining the practicality of the derived units. My own assessment of whether the SI unit for a physical quantity is either readily *visualized* as the product of base units or is *practical* is indicated in columns V and P respectively.

Range of Quantities which have Dimension $\langle\,\text{time}\,\rangle = 0$.

SI: V	P	Physical Quantity	Extremes		Natural Construct
Yes	Yes	Velocity (meter per second=N8.5)	crawl speed of sound in air	N10 N6.0	speed of light=1=P0 speed of sound=c=(bulk modulus/density)$^{1/2}$=(for solids) $(e^2/2\pi d_o m_C)^{1/2}$=N4.3
No	No	Specific energy (joule per kilogram=N17.0) Latent heat of vaporization: Heat value of fuels:	helium water lignite natural gas	N12.7 N10.3 N9.5 N8.5	R_∞/m_C=N8.9
No	No	Permittivity (farad per meter=P11.1)	vacuum barium titanate	P0 P5	permittivity of free space=ϵ_o=P0
No	No	Permeability (henry per meter=P5.9)	vacuum supermalloy	P0 P6	permeability of free space=μ_o=P0
No	No	Impedance (ohm=N2.6)	1 milliohm 10 megohms	N5.6 P4.4	impedance of free space=Z_o=P0
No	Yes	Charge (coulomb=P17.8)	electron charge coulomb	N0.9 P17.8	$e^2/2=\alpha=1/137$ $N_A^{2/3}\,e$=P15.0
No	No	Magnetic flux (weber=P15.3)	quantum of flux weber	P0.6 P15.3	$\Phi_o=1/2e=4$=P0.6 flux from pole of ferromagnet $=(\Delta m/\Delta V)A=(\mu_B/d_o^3)(N_A^{2/3}d_o^2)=$ P12.2
No	No	Heat Capacity	atom of He gas	P0.2	$dE/dT=\tfrac{1}{2}$ per degree of freedom

Range of Quantities which have Dimension $\langle \text{time} \rangle = -1$

SI: V	P	Physical Quantity	Extremes		Natural Construct
Yes	Yes	Acceleration (meter per sq. second = N3.5)	g/100 10 g	N4.5 N1.5	g = acceleration of gravity = N2.5 (v^2/r for an electron in hydrogen $= \alpha^2/a_o$ = P19.8!)
Yes	Yes	Frequency (hertz = N4.9)	pulse beat Cs atomic beam resonator	P5.0 P14.8	$R_\infty = me^4/8$ = P20.5
Yes	Yes	Temperature (kelvin = P15.3)	1 kelvin 10,000 kelvins	P15.3 P19.3	$R_\infty (m/m_C)^{1/2}$ = P18.4 T_{atmos} = P17.7
No	Yes	Voltage (volt = P20.2)	10 microvolts 100 kilovolts	P15.2 P25.2	R_∞/e = P21.4
Yes	Yes	Current (ampere = P22.8)	10 nanoamperes 500 amperes	P14.8 P25.5	$R_\infty/(eZ_o)$ = P21.4
Yes	No	Energy (joule = P38.1)	joule ton of TNT	P38.1 P47.7	R_∞ = P20.5 $N_A R_\infty$ = P44.3
Yes	Yes	Mass (kilogram = P55.1)	gram ton	P52.1 P58.1	mass of carbon atom = m_C = P29.4 $N_A m_C$ = P53.2

Range of Quantities which have Dimension $\langle$ time $\rangle$ = +1

SI: V	P	Physical Quantity	Extremes		Natural Construct
Yes	Yes	Time (second=N4.9)	microsecond	N10.9	mean solar day = 1 = P0
			year	P2.6	
Yes	Yes	Length (meter=N13.4)	micrometer	N19.4	$d_o = 2a_o = N23.4$
			mile	N10.2	$N_A^{1/3}d_o = N15.5$
No	No	Capacitance (farad=N2.3)	3 picofarads	N13.9	$N_A^{1/3}d_o = N15.5$
			0.3 farads	N2.9	
No	Yes	Inductance (henry=N7.5)	microhenry	N13.5	$N_A^{1/3}d_o = N15.5$
			500 henrys	N4.8	
Yes	No	Coefficient of thermal expansion (inverse kelvin = N15.3)	invar	N21.3	$R_\infty^{-1} = N20.5$
			paraffin	N19.3	
No	Yes	Specific heat capacity (joule per kilogram kelvin = N32.2)	lead	N30.0	$m_C^{-1} = N29.4$
			H_2 gas	N28.0	

Range of Quantities which have Dimension < |time| > = 2

SI: <t>	V	P	Physical Quantity	Extremes		Natural Construct
+2	Yes	Yes	Area (square meter = N26.8)	sq. millimeter	N32.8	d_o^2 = N46.8
				sq. kilometer	N20.8	$N_A^{2/3}d_o^2$ = N31.0
−2	No		Electric field (E) (volt per meter = P33.7)	at earth's surf.	P35.7	Electric field of an ion at neighboring nucleus of ionic solid $E_c = e/4\pi d_o^2$ = P44.8
				breakdown of dielectrics	P42.7	
−2	No		Electric flux density (D) (coulomb per sq. meter = P44.6)	same as for E		same as for E
				same as for E		
−2	No	Yes	Magnetic field (H) (ampere per meter = P36.2)	ampere per meter	P36.2	same as for B
				coercive force of silmanol	P41.9	
−2	No		Magnetic flux density (B) (tesla = P42.1)	at earth's surf.	P37.9	saturation B field in iron =
				saturation field in iron	P42.4	μ_B/d_o^3 = P43.2
−2	No	Yes	Thermal conductivity (watt per meter kelvin = P41.2)	air	P39.6	(1/3) × concentration × speed of sound × mean free path = (for solids)
				silver	P43.8	$(1/3)d_o^{-3}cd_o$ = P42.0
−2	Yes	Yes	Power (watt = P43.0)	watch	P37	power of explosions = $(R_\infty c/d_o)(N_A^{2/3})$ = P55.5
				ship	P51	
−2	Yes	Yes	Force (newton = P51.5)	gram weight	P49.5	breaking strength of solids = $(eE_c)(N_A^{2/3})$ = P59.8
				ton weight	P55.5	

Range of Quantities which have Dimension $< |\text{time}| > = 3$

SI: $<t>$	V	P	Physical Quantity	Extremes		Natural Construct
+2	Yes	No	Volume (cubic meter=N40.2)	cubic centimeter cubic meter	N46.2 N40.2	d_o^3=N70.2 $N_A d_o^3$=N46.4
−3	Yes	No	Current density (ampere per sq. meter=P49.6)	in household wire in skin of superconductor	P55.6 P60.0	charge density of conduction electrons × velocity of sound=$(e/d_o^3)(c)$=P65.0
−3	Yes	Yes	Charge density (coulomb per cubic meter=P58.0)	semiconductor (Si) copper	P55.1 P68.8	charge density of conduction electrons in a metal=e/d_o^3=P69.3
−3	Yes	No	Concentration (number per cubic meter=P40.2)	air diamond	P65.8 P69.5	density of atoms in a solid $=1/d_o^3$=P70.2
−3	No	No	Surface tension (newton per meter=P64.9)	gasoline mercury	P62.7 P64.6	gradient of force between ions $=2e^2/4\pi d_o^3$=P67.0
−3	No	No	Dynamic viscosity (pascal second=P73.4)	air glycerine	P68.7 P73.4	(1/3) × density × speed of sound × mean free path=(for liquids) $(1/3)(m_C/d_o^3)c d_o$=P71.4

Range of Quantities which have Dimension $< |\text{time}| > = 4$

SI: $<t>$ V P	Physical Quantity	Extremes		Natural Construct
-4 No No	Irradiance (watt per square meter=P69.9)	blackbody at temp. of 300 K sun's surface	P72.4 P77.6	Stefan-Boltzmann law for irradiance of blackbody= $(2\pi^5/15)T^4 = 41\,R_\infty{}^4 = P83.6$
-4 No No	Pressure (modulus of elasticity) (pascal=P78.4)	threshold for human ear modulus for piano wire	P73.7 P89.7	atmospheric pressure=P83.4 Young's modulus for a solid = gradient of force at atomic distance per atomic distance= $(2e^2)/(4\pi d_o{}^4)=P90.4$
-4 Yes No	Density (kilogram per cubic meter=P95.3)	air osmium	P95.4 P99.7	density of a solid= $m_C/d_o{}^3=P99.6$
+4 Yes No	Specific volume (cubic meter per kilogram=N95.3)	osmium air	N99.7 N95.4	reciprocal of density= $d_o{}^3/m_C=N99.6$

Appendix II to Chapter 4

Petition Concerning Metric Conversion
To the Colorado Delegation to Congress

Many now believe that eventual acceptance of the International System of Units (the proposed standardized metric system) is "inevitable." *If* this is true, then we believe that the "inevitable" conversion should take place naturally *without* the prodding of the ten-year program now being considered by the Congress.

As scientists, we are well aware that the International System of Units is but a step in the evolving relationship of man to nature. Its origins are in the metric system, originally formulated during the French Revolution. The framers of that system could not foresee the impact which subsequent developments would have on technology. Units for such developing sciences as electricity or optics were first introduced *ad hoc*. Then, early in this century, the metric system itself was enlarged to encompass these nineteenth century developments. With little further modification, this enlarged system has become the International System of Units.

Although recognized by statute the world over, the International System has not been fully accepted—even by European scientists and engineers. Deprecated units for such common concepts as force, pressure, and magnetic field persist because they are more convenient than their counterparts in the International System. More importantly, as a crystallization of basically nineteenth century technology, this system is poorly suited to twentieth century developments. These developments have been in our understanding of fundamental atomic and molecular processes and in the use of binary—rather than decimal—arithmetic in computers.

Within a generation, man may well devise a truly modern system—one which combines the coherence of the International System with the convenience of our customary one. We are concerned lest a legislated conversion to a rigid system deprive future generations of the benefits of a truly optimal system of measurement.

Therefore, be it resolved that we, twenty-two members of the faculty of the Department of Physics and Astrophysics at the University of Colorado, urge the rejection of pending metric conversion legislation.

5.
Lessons for the Metric System: Decimal Time

Richard A. Carrigan, Jr.
Physicist and Assistant Head of Research Division,
Fermi National Accelerator Laboratory

Universals

Permeating all cultures are systems of standards that govern the existence of societies within the cultures. Languages exist so people can communicate with one another. A language may vary from place to place but there will always be some grammar and a degree of style to which all the participants adhere. Since the dawn of civilization, alphabets have come and gone with some of them now pervading many different countries throughout the world. Occasionally an old language, such as Latin, is still in use with no modern society to claim it. Systems of laws and rules are developed by societies which are often quite arbitrary but still function as signposts of behavior. In another corner of the house of rules are standards of weights and measures.

How these languages, signs, weights and measures, laws and other systems to regulate social intercourse came into existence, diversified, then merged together to become more homogeneous is an intriguing topic. Just as interesting is the question of how much uniformity in all these systems is desirable. For example, would it be useful to have a universal language that permeated the whole world? Or is the Tower of Babel that is the world today something that adds diversity and interest to modern civilization?

On the face of it, the diversity in language is a real problem for the traveler. In recent times a need has arisen for systems of traffic signs that transcend a national language. Such a system has been in use for many years in Europe, with the signs spreading across the continent through many different language regions. But language still enters nearly every other activity of the traveler. An example is the requirements of airline transportation. Indeed, English is ordinarily adopted among the airline pilots and flight controllers of the world as the language of choice to overcome some of these problems.

An improvement in consistency in one area may bring about problems in another. International traffic signs represent just such a case. At present there are two widely used systems of symbols for representing language: Alphabetical systems, which basically represent sounds of parts of words, and calligraphic systems such as Chinese that represent words with characters. On the face of it, alphabetic systems seem to be the most practical for ordinary written text. A limited set of characters is able to represent an enormous number of words. This makes printing books alphabetically much easier than assembling type in calligraphy. In adopting international road signs, an element of calligraphy is introduced into western culture. Some reflections on this show that more and more such symbols are being used. Is there some risk that eventually the stick sign for men and women will somehow enter written languages and replace the alphabetic words? Questions of this type are quite serious and appear to have been given little consideration.

Most of the systems of organization that have come down to modern civilization grew out of practical problems. A rough grammar may exist along with a sturdy vocabulary, but much more often than not these systems of organization have grown up empirically. An easy example is the ordering of the letters of the alphabet. Before children in school can do much else they must learn the alphabet in a certain order, A, B, C, . . . etc. Without that they cannot even look up a word in the dictionary. But on the face of it there seems to be no obvious reason to order the letters as they are. With numbers, of course, this is not the case, and a logical system obviously exists.

There are, however, some systems that have come into being with quite logical grammars and even somewhat consistent vocabularies. The metric system is an example. The metric system was given to history by a band of enlightened radicals who set about changing the very foundations of many different aspects of their culture during the French Revolution.[1] Many of the fallouts from the revolution were bitter indeed. Two hundred years later scientists still remember that Lavoisier lost his head in the scramble. But the "rationality" that led to the guillotine also created a system of weights and measures that at least made sense within itself. There are, of course, other earlier examples of logical systems winning out over less logical systems. One example is the introduction of the decimal system to replace Roman numerals.

For some years I have been interested in the question of how such systems evolve. I have wondered if international languages could be introduced into the culture, hopefully without the drastic measures taken in the French Revolution. I have also wondered whether such an introduction would make sense. When this question is considered, one soon begins to look for case studies of the introduction of actual standards into societies to see how well some succeeded and others failed. The metric system of weights and measures is a beautiful example. Over the last two centuries, the units derived from the metric system have come into wide use, but not without a great deal of difficulty along the way.

It is not widely recognized that, at the same time the metric system was devised, the French Revolutionaries also proposed to revise the system of time-keeping.[2] During the Revolution, the French tentatively adopted a system of decimal time along with a new calendar. The decimal time system fared poorly, and disappeared from the scene within a few short years. The comparison of the history of these two changes of measurement standards can give some insight into how systems of standards and of language organization can change.

Decimal Time

By design, the metric units for weights and measures are quite consistent with the decimal system, but this is not the case for ordinary time units. Time is logically divided into two parts: the subdivisions of a day, and the calendar within which the days fall.

The calendar is a problem. Outside of drastic celestial engineering there appears little that can be done to circumvent the facts that the earth revolves 365¼ times a year, there are four seasons, the moon revolves around the earth once every twenty-nine days or roughly twelve times a year, and that work periods have habitually been divided into seven-day weeks or the equivalent of a phase of the moon. Almost from the dawn of civilization societies have struggled with this problem of the odd and fractional number of days in the year because of the practical need to keep accurate planting times for agricultural purposes. It has been a long struggle involving drastic calendar changes as calendars fell out of synchronism and no longer performed their proper function. During the French Revolution a calendar was devised, still with twelve months, but with each month containing three weeks of ten days each.[3] The five remaining days were thrown into a holiday period called the "Sans Culottides."

For the time division within the day the present clock divides each day into twenty-four hours, twelve each before and after noon. Each of these hours is divided into sixty minutes. The name "minute" derives from the Latin *minuta*, that is, small. In turn, minutes are divided into sixty seconds where "second" literally means "second minute." It is this second that forms the basic unit of time in the metric system. Now, it is certainly a cumbersome set of units and one that is far from the decimal system that leads logically from the fundamental practical standard in this case—one day—down to the actual unit, one second. The units are not even consistently sexidecimal, but rather a mixture of twenty-four hours and sixty minutes and sixty seconds. In addition, the most widely used nomenclature for numbering the hours, that is the A.M.–P.M. convention, leads to a great deal of confusion. It is difficult for most people to properly tell the number of hours that have transpired between a time in the afternoon and a time in the morning using that convention. Further, there is no logical consistency between the names for the three fundamental units—days, hours, and seconds—in any of the modern languages. This situation can only be described as chaos. The present SI metric system avoids the problem in the sense that the second is defined as the standard unit.[4] Minutes, hours and days are allowed words in the metric vocabulary with some recognition that they are entirely inconsistent with the normal metric system.

The French godfathers of the metric system realized this when the metric system was introduced during the revolution. They proposed a new system of time-keeping in which the day was divided into units of ten decimal hours. These units were further subdivided by tens to form smaller and smaller units. For example, the equivalent of a decimal second would correspond to one part in a hundred thousand of a normal day and 0.864 of an ordinary second. The Committee of Public Instruction, an adjunct of the Revolutionary National Convention, undertook a thoroughgoing examination of the question of decimal time. Ultimately, the review touched on many aspects of what a useful measurement system should have.[5] A subcommittee from the Committee asked scientists to examine the very heart of time-keeping systems, with the possibility of proposing different forms of watches and clocks. Simple decimal clocks, watches and even sundials were put into operation during the height of the Revolution. Figure 1 shows a beautiful example of a double dial watch from the French Revolution with one dial showing decimal time and the other showing conventional time, while Figure 2 shows an interesting decimal sundial.

It should be noted that these considerations were not side-issues in the Revolution but quite at the heart of the revolutionary dialectic. In a sense they epitomized the determination of the Jacobins to execute an entirely rational revolution. As the Revolution moved inexorably on, both the metric system and decimal time were tentatively adopted.

As the reign of terror raged, however, a few cool heads continued to deliberate further on these matters. In the spring of 1795 Prieur of Cote d'Or presented the general case for the adoption of a metric system of weights and measures to the National Convention. He did not press for the adoption of decimal time, instead suggesting that initially it might be more useful in technical areas and then come into more general use later. Ultimately, the

Figure 1. On this exquisite example of a double dial watch from the French Revolution, Liberté, or La France, turns her back on conventional time and looks expectantly toward the dial representing the new decimal time-keeping system. The time is 9:58 p.m. or roughly nine-tenths of the way through the day. From the collection of Au Vieux Cadran, Reprinted from S. Guye and H. Michel, Time and Space: Measuring Instruments from the 15th to the 19th Century *(New York: Praeger, 1970).*

Figure 2. Lovely decimal sundial from the French Revolution. Normal time is given in outmoded Roman numerals, decimal time in Arabic numbers. Reprinted with the permission of the Boston Museum of Fine Arts.

National Convention approved the metric system while suspending the adoption of decimal time indefinitely.

One of the central questions about decimal time is why the
system failed so totally in the French Revolution. M.J. de Rey-
Pailhade, a keen advocate of decimal time around 1900, suggested
several reasons:[6] 1. Decimal calculations were little understood at
the time of the introduction of the new system; 2. serious modifications were required for existing time-keeping devices; 3. the
absence of new terms for the new units that were introduced. In
his Harvard dissertation on the French Revolutionary calendar, J.
Friguglietti cites Prieur of Cote d'Or as giving five official reasons
for suspending the adoption of decimal time when the metric system was sanctioned in 1795: "1. It would offer no marked advantage over the old system and would 'throw disfavor' over the new
system of measures as well; 2. the strength of habit would make
enforcement impossible if indeed such regulations were lawful; 3.
popular confusion would result; 4. the expense of changing all
time pieces would be enormous; and, 5. citizens would intensely
dislike converting their watches to the new system and watch-
makers would be unable to dispose of their old stock." Friguglietti
himself suggests that the reason was political rather than practical,
stating, "The decision of the Convention to abandon decimalized
time reflects the growing reaction that followed the disintegration
of the revolutionary government after 9 Thermidor.* Decimal
time joined the maximum† on the scrapheap of terrorist
measures."

---

*The French revolutionary calendar began on September 22, 1792, the day
after the abolition of the monarchy. It consisted of twelve months, each of
thirty days, with names that corresponded to the seasons of the Northern
Hemisphere: Vendémiaire (month of vintage), Brumaire (month of fog), Fri-
maire (month of frost), Nivôse (month of snow), Pluviôse (month of rain),
Ventôse (month of wind), Germinal (month of budding), Floréal (month of
flowers), Prairial (month of meadows), Messidor (month of harvest), Thermidor (month of heat), and Fructidor (month of fruit). 9 Thermidor – July 27,
1794 – was the fall of Robespierre and his subsequent execution by guillotine.
†The maximum was an unpopular and unsuccessful price control system used
during the Revolution.

These reasons all seem quite plausible. My own view is that other political considerations may have also played a part. Prieur, perhaps an expedient compromiser, had the type of tool a politician dearly loves. His proposal had two parts, the metric system and decimal time. He was able to sacrifice one part, decimal time, to gain acceptance of the metric system.

In the nearly two hundred years since the Revolution decimal time has been reinvented many times. M.J. de Rey-Pailhade mounted a determined forty-year campaign in France around the turn of the century complete with decimal time yacht races, proposed law changes and more decimal clocks. His efforts were quite without success.

From the practical side, decimal time is not without problems. One serious difficulty is that natural activities now divide conveniently into three periods: work, leisure and sleep; a system of ten decimal hours is difficult to divide into three equal parts. Another obvious problem with a decimal time system is that the standard unit of time would change. This would change the scientific units throughout. Velocities, for example, would have to be quoted in terms of a decimal second.

My own interest in decimal time is centered principally around the history of the concept and what the roots of its failure were in the face of metric success. Nevertheless, I have sometimes wondered whether one might advocate introducing a decimal time system in the spirit that Rey-Pailhade did. Personally, I could not mount such a campaign. On the other hand, it seems to me that it is necessary to maintain a degree of open mindedness and awareness of the possibilities of decimal time. In some future, the imperative of decimalization might mandate its adoption. It is questions of this sort that are really most at issue in trying to understand the U.S. view of the metric system and the future evolution of the metric system for the world as a whole.

Overall, some of the ingredients that killed decimal time still choke out widespread adoption of the metric system in the United States. Poor understanding of the units, strength of habit, confusion, and dislike of conversion of equipment are still problems. And real practical and political considerations may set the tone of acceptance. These lessons, known from decimal time, are helpful in viewing the metric system.

Metric Units

The present international system of metric units is a largely logical and sound structure. By adhering closely to the decimal system, irregular arithmetic operations are avoided. Some degree of notational consistency is maintained. All other points being equal, it is much better than the United States-British system of feet and pounds, where mixed arithmetic and ad hoc names run rampant.

Before discussing the problems that beset the metric system at present, one further point about systems of weights and measures should be emphasized. There are two distinct features to a system of standards. The first is the vocabulary and grammar associated with that system. In the case of the metric system this means the choice of decimal units as a grammar and the choice of consistent decimalized names as a vocabulary. The second feature of any system of weights and measures is the standards themselves. Some reference object, be it a gold block or an atom, must weigh so many kilograms. The second point is extraordinarily important, both from the standpoint of the man in the street with his interest in securing proper value for his money, and for the scientist who must make consistent measurements the world over. The present SI system of metric units is handled very well from the standpoint of standards of measurements. The definition for length is now in terms of a fundamental atomic measurement that appears to be immutable and universal. There is some hope that the definition for mass may eventually be fixed in a similar manner. No doubt these standard conventions can be improved, but from a practical point the definitions are both reasonable and reflect the current state of the art. This situation owes itself to the dedicated work of many people in the national standards organizations of a number of countries. The vocabulary and grammar of the metric system, however, may not be in such good shape.

There are a number of difficulties that exist with the metric system. In this regard I will quote some of my own favorite problems with the metric system rather than giving a comprehensive criticism.

The SI metric system adopts a temperature scale, Celsius, the old degree centigrade, that is rather arbitrary. From a scientific standpoint, it would be more comfortable to have this unit grounded on a more fundamental basis than the property of water. Worse, from a scientific standpoint, is the failure to go directly to the more technically consistent scale of degrees Kelvin.

The present metric system adopts a unit for magnetic field called the tesla. There seem to be few people in the world who like that unit. The present units of choice are the gauss and kilogauss. Units of electric charge are defined in terms of electrical current. From an intuitive standpoint, it would appear to be more sensible to define a static quantity such as electric charge and let electric current be a derived quantity. Either approach is satisfactory from the standpoint of physics and either is probably correct from a didactic standpoint. The conceptual problem with a fundamental definition in terms of a current can be seen by an example. Instead of defining mass, the metric system could have called momentum a fundamental unit. But measurement of momentum requires some sort of dynamical process. Astronauts measured their mass by a centrifugal system, but a bathroom scale is far more convenient for earthbound people.

In the metric system, car velocities, the velocities most people understand, are given in suspect units in metric countries. They are normally quoted in kilometers per hour for rather obvious reasons. The correct SI unit would be meters/second.

The present metric system adopts a grammar in which units are defined in terms of divisors and multipliers of a thousand, that is, milligrams, grams, and kilograms. One may wonder where this factor of a thousand came from. Why not a factor of a hundred instead? The factor of one thousand leads to some fairly unusual combinations. Locomotives in France are said to be designed in millimeters. The first metric unit that many people were exposed to, the centimeter, is by this rule banished. Another difficulty in this combination of vocabulary and grammar is the fact that the letters and prefixes chosen for metric terminology are quite inconsistent. An example I have faced is the need for a metric prefix for times on the nuclear scale, that is, 10^{-24} seconds. Readily available metric nomenclature tables go only to the prefix atto for 10^{-18}. In a sensible system, the nomenclature for the next steps would be self-evident. The use of kilogram, rather than gram, for the basic unit of mass builds in a whole set of special rules for mass nomenclature. Strict adherence to logical rules would designate the gram as a millikilogram. Indeed, the whole question of names within the metric system is a sore point. Most of the names, in the words of David Bartlett,[7] an incisive metric critic, are taken from the names of dead physicists. Perhaps a better approach would have been to use some logical structure that would have been alphabetized, or to let the units suggest the quantity they represented.

Some metric critics such as Kenneth Boulding[8] have suggested that the metric system is not human in scale. My own view is that this is not a particularly telling criticism. The difficulty comes with defining a human scale. A nuclear physicist feels quite at home with units that are one million billionth of a meter while an astronomer is almost closed in with units of a million billion kilometers.

One final criticism of the metric system is often heard. This is a charge against the very number system itself, the decimal system. The argument usually holds that some form of a binary system, for example, octal or sexidecimal, would be more appropriate because this system is widely used in computers. Occasionally a number system based on twelves is suggested because of the number of common divisors available. In that regard it is sometimes remarked that the present system of time-keeping is more perfect than the system of weights and measurements. This is nonsense, for the system of time-keeping doesn't go by any uniform system of twelves.

The decimal system of numbers is very pervasive, so that even in computer systems where binary and octal systems are frequently used, the intermixing of the number systems causes inordinate difficulties. One of these difficulties has to do with the failure to distinguish the two types of numbers by different symbols, so that a '17' in the octal system is different than a '17' in the decimal system. Now when one uses '17' in computer systems it is often not possible to be sure whether it is an octal '17' or a decimal '17.' (To be sure, conventions are sometimes adopted to permit a distinction between number systems.) One often has to determine which number system is being used by a process of elimination, such as noting that there are no eights and nines present. Serious revolutionaries may argue that there is no time like the present to get the matter of the number system straightened out. They believe that by ignoring the current dilemma we may lay the foundation for even worse problems 200 or 500 years from now. Of course, the abandonment of the number system would also imply the abandonment of the present metric system.

To me, perhaps the most difficult aspect of the metric system is understanding how the system will evolve. It is certainly true that the metric system has changed since its inception nearly 200 years ago. I learned a metric system called the CGS, centimeters-grams-seconds, about two decades ago. By the time I started to teach physics myself, that system had changed to an MKS system (the SI system), that is, meters-kilograms-seconds. At the same time all of the laws of physics pertaining to electricity and magnetism had changed. Nevertheless, when we ask how the metric system will move from its present form into a future metamorphosis, the answer is not so clear. As near as I can determine, these changes are brought about by meetings of international bodies, responsible in large part to scientific organizations. However, it should be emphatically stressed that the system of weights and measures is not the property of scientists or standards agencies alone.

But this eclectic list of metric system deficiencies does little to diminish the manifold virtues of a widespread adoption of the metric system. Rather it shows that the system should continue to change and improve. Again, it should be stressed that the metric system with its uniform use of decimal notation and reasonably consistent nomenclature is the best widely available measurement system in the world today.

One important aspect of metrication that is often overlooked is the question of open versus closed systems. For large scale industrial production it is relatively easy to adopt some special fasteners, units or standards. These systems can be viewed as closed. For example, automobile manufacturing and servicing is closed since nearly all of the parts can be supplied by the manufacturer and sent to the distributors. On the other hand, United States building construction materials are open systems in which the standard of sixteen-inch spacing for joists has far-reaching impacts on a great variety of goods. Indeed, that standard may be the hardest one to shed with metrication since replacement units might be needed from thirty to fifty years in the future.

A fascinating example of a closed system that has expanded into an open system is the modern Dual Inline Pin or DIP socket used for many integrated circuits. Figure 3 shows a sixteen-pin DIP socket along with an accompanying integrated circuit chip. Pin spacings are exactly 0.1 inches rather than 2 or 2.5 millimeters. Note that this standard was set after some of the virtues of metrication were very clearly known. These sockets, introduced about a decade ago, revolutionized electronic manufacture by providing an extremely simple architecture for printed circuit boards. A major fraction of the world's electronic production now uses these devices. Up to now the system has existed without serious impact on other technologies. However, as microelectronics has made its way into even more aspects of modern life the 0.1-inch standard could affect other areas of manufacturing. Clearly the DIP socket was a casually adopted standard that later had an enormous impact.

Several lessons about metrication can be learned from the case of the DIP socket and a recognition that some systems are more open than others. For a truly closed system there is clearly plenty of room for choice. Manufacturers, such as automobile producers, should adapt this strategy to the market. They need not have to rush into metric except where it does them good. If they feel that metrication widens their markets they may proceed faster. For open systems and new standards the United States should try hard

Figure 3. DIP socket used for integrated circuits. Pin spacings are exactly 0.1 inch on these modern devices. The standard has become almost as endemic in electronic architecture as the sixteen-inch joist spacing is in the building trades.

to adhere to the metric system. However, it is important to remember that the United States also has something to offer. As the last group on the bus the nation can adopt even more consistent systems that replace outmoded or poorly conceived standards.

A case could be made that the United States has approached metrication with too idealistic a perspective. Perhaps there has been some of the fervor that the decimal time advocates exhibited during the French Revolution, when logic overrode for a time the possibility of popular confusion and the impact of conversion costs. A Prieur may need to come forward to speak for moderation. Pro-metric arguments claim that the system will give the nation wider markets, simplify the citizens' lives and be rational. Another viewpoint is that European pressure for total metrication in the common market moves the United States out of trading areas and gives the Europeans and Japanese a trading edge. The ready acceptance of metrication by United States standards agencies and the government rather than an aggressive move to keep goods made with old units in place may have lost the United States a bargaining chip.

Perhaps the United States should have followed an approach requiring a market place quid pro quo in return for United States metrication. Several possibilities come to mind. The United States might have advocated English as the universal language for instructions. A second possibility would have been mandating the use of simplified universal fasteners. The story of the United States proposal for a simplified set of universal metric fasteners has been reviewed by Stanley Mallen.[9] In essence, United States standards groups proposed a simplified and improved set of metric fasteners several years ago. In spite of the obvious advantages to inventory, European manufacturers would not adopt them. Perhaps the government should have worked hard for adoption of those standards as a basis for the United States adoption of the metric system. It might even be that the country could still use this set of fasteners with the hope that their obvious advantages would help the nation to leapfrog European and Japanese goods in the market.

In summary, it is important to remember that change will be the rule rather than the exception. With so many problems and with new types of manufacturing, technology and science coming along, it is important to always be prepared to modify and improve all standards systems in the future. The metric system is good, but better approaches may come along. With temperate and careful consideration these approaches will further foster international commerce and understanding.

Notes

1. See, for example, G. Bigourdan, *Le Système Métrique des Poids et Mesures* (Paris: Gauthiere-Villars, 1901).

2. Richard A. Carrigan, Jr., "Decimal Time," *American Scientist* 66 (1978): 305.

3. J. Friguglietti, "The Social and Religious Consequences of the French Revolutionary Calendar" (Ph.D. diss., Harvard University, 1966), p. 23.

4. "International System of Units," *Federal Register* 40 (1975): 25837.

5. J. Guillaume, ed., *Procès-verbaux du Comité d'Instruction Publique de la Convention Nationale*, 6 vols. (Paris: J. Guillaume, 1890-1907), 2: 882.

6. M.J. de Rey-Pailhade, *Le Temps Décimal* (Paris, 1894).

7. See chapter 4 of this volume.

8. Chapter 3.

9. Chapter 2.

6.
The Metric System:
Costs vs. Benefits:
A Summary Overview

David T. Goldman
Associate Director for Planning,
National Measurement Laboratory,
National Bureau of Standards

The contributors to this volume have presented five different aspects of the present public and industrial concern over the desirability of adopting the metric system in the United States. Covered here has been the spectrum of the controversy, ranging from the case for the immediate adoption of the internationally approved International System of Units (SI) version of the metric system to a plea for examining alternatives to this system before its adoption precludes the acceptance of a superior system of units. In spite of my ingrained cultural bias towards the customary system of units, I believe that it is both desirable and inevitable that the United States facilitate the adoption of the SI units for the purpose of communication of the magnitude of measurement quantities in industrial and commercial transactions. Use in business and industry will be the driving force for the ultimate transition from customary units to SI in everyday language.

The 200-year development of the metric system of units as a rational language system for measurement to replace the plethora of individual regional or national systems has been summarized many times.[1] The early history of the United States contains instances when total conversion to the decimally-based metric system was considered and rejected even though our primarily decimally-based system of coinage was established in order to take advantage of the arithmetic simplicity of any system based on powers of ten.[2]

At first the metric system was proposed to replace the measurement units for those quantities most familiar to all through normal commodity or property exchange—weight and length and volume—or "weights and measures" in common parlance. The development of technology and the dissemination of its benefits across national boundaries in the nineteenth century demonstrated the desirability of putting all measurement systems on a common basis. As a result of the "Convention of the Meter" in 1875, with the United States one of seventeen signers, an International Bureau of Weights and Measures (BIPM)* was established to maintain custody of the international prototype standards of the meter and the kilogram which were then the material basis for the metric system. An International Committee for Weights and Measures (CIPM)* to coordinate activities of the signers of the Convention and a General Conference on Weights and Measures (CGPM)* for general oversight of the BIPM and the development of the metric system were formed.[3] Over the next century, five additional units were added to the meter and kilogram to form a consistent system built upon these seven base units from which all other units could in principle be derived. The added units of time, electric current, temperature, amount of substance and luminous intensity with the original units for length and mass form the basis for the subset of the metric system which was officially designated by the CGPM as the "International System of Units" or SI. In 1964 a subcommittee of CIPM, entitled the Consultative Committee on Units (CCU), was established to advise the CIPM on numerous proposals that are made to improve SI. I presently serve as the representative of the National Bureau of Standards to the CCU.

By 1960 much of the present SI had been formally established. During the next fifteen years, selected derived units were given special names (e.g., the pascal, the SI unit of pressure), and units which could be used together with SI units were identified. This system is still evolving as is evidenced by the recent recommendation by CCU to adopt as an SI unit the sievert, symbol Sv, a

---

*These abbreviations are taken from the first letters of the French versions of the names of these organizations.

special name for the unit of the quantity "dose equivalent," important in the field of radiation protection.[4] Upon approval by the parent committee, CIPM (in September 1978), and adoption by CGPM (considered in October 1979), the SI will be modified accordingly. A document summarizing the latest status of the SI is published and revised as needed by BIPM, with the assistance of CCU. The latest version is the 3rd edition, published in 1977 and available in English translation (see appendix).[5] The SI as embodied in this document forms the basis for recommended practice with varying degrees of required compliance, depending upon the technical area or use of the system.[6] A voluntary consensus in certain fields of technology may result in a decision to use, for a limited time, units outside the SI.[7] Changes in the system can be proposed for consideration by writing to the Director of BIPM, though the deliberated recommendations of international technical organizations are clearly given primary attention.

I have summarized the development of the International System of Units because I feel that it is important that an internationally agreed upon system of units be the language of communication for transactions across national boundaries. A choice of the best system of units cannot be made upon purely technical grounds. Different systems of measurement evolved in different cultures, and there is a pleasing familiarity in describing a city as a "mile high," or a piece of lumber as a "two by four." Professor Boulding has raised very interesting questions concerning the practical nature of the most significant SI units. It seems likely that the origin of most of the common units was rooted in human experience, giving rise to many measurement language systems for different users. H. Arthur Klein, in a section of his book *The World of Measurement*[8] entitled "The Babel Behind Us," has identified for the same physical quantity many units which are not related to each other by any rational numerical ratio.

The confusing multiplicity of units meant that a proposed single system of units could find a receptive audience among the commercial interests of the world. Though the metric system did not find many adherents during the first century after its invention, by the start of the second century it was found that international commerce and trade, even on its limited scale compared to today, could be improved by the use of a single system of units that transcended national boundaries. I agree with Professor Boulding that the choice of the base unit of length, the meter, defined as one ten-millionth of the length of a quadrant of the earth's meridian, and the application of simple rules for the derivation of new units, has resulted in an international system in which the units are of a magnitude such that without a prefix they are often not of appropriate size.[9] Thus, the unit of pressure, the pascal, is defined as one newton per meter-squared, an extremely small unit. A practical unit of atmospheric pressure, the bar, is equal to 100 kilopascals. It appears that the widespread use of the bar in meteorology will result in its continued use for the foreseeable future, a fact recognized at the most recent CCU meeting.[10]

To summarize, the SI is a system of units comprised of seven base units and two supplementary units (the radian and steradian), units formed by combining the base and supplementary units, and a set of prefixes to form decimal multiples (or submultiples) of these units. The system is coherent; that is, a unit for every physical quantity is available, derived in the simplest possible manner from the base and supplementary units without the introduction of any factor other than the factor one.

Furthermore, there exists a mechanism for modification of the system. However, the major revision proposed by Professor Bartlett, while satisfying to the scientist who must be concerned with many orders of magnitude in the range of physical quantities, can hardly be expected to appeal to our industrial complex or to the general public.[11] Examples of important changes in SI are the addition of the mole as a base unit in 1971 and the steps now in progress, as described above, to recognize the sievert as the SI unit of dose equivalent.

On the other hand, Professor Bartlett's suggestion of using a single base quantity, duration of time in this case, does have an intellectual appeal while not really changing the use of SI in practical measurements. The measurement of the velocity of light to a much higher accuracy than ever before will make it possible to replace one or the other of the base units of time and length; e.g., the unit of length can be defined in terms of the distance traveled by light in a specified period of time.

The attractive features of the metric system have led to its almost universal adoption. It has long served as the language for scientific communication in the United States. The industrial sector, citing economic and trade reasons, is in the process of converting. Mr. Mallen has pointed out the advantages derived from increased use of the metric system in product reliability and cost reduction.[12] Whereas a preference for SI might be obvious for a technology-intensive industry dominated by a few large multinational firms, it was interesting to learn of the advantages available to the fragmented construction industry in converting to the metric system in an article entitled, "Our Metric Future."[13]

I firmly believe that selective conversion now will result in cost avoidance later and that encouragement of voluntary metric conversion is consistent with the intent of Congress[14] and the views of the present Administration.[15] As the private sector recognizes the benefits of conversion, and specific sectors adopt the metric system, it will become all the more desirable for the remaining sectors to be on a consistent measurement basis with those who have already recognized the advantages of conversion. In other words, I believe that the successful conversion to metric by certain industries will be the driving force behind ultimate voluntary conversion to the International System. This process has already started and it is the government's responsibility through the United States Metric Board to coordinate this conversion in the least costly manner possible.

In view of the desirability of an orderly transition to the metric system of units, there are certain warning signals which must be raised. A very recent General Accounting Office report questions both the desirability and public acceptance of conversion.[16] This report has been widely referenced by journalists opposed to changing from the inch-pound system. An increase in the unit price of liquor with the switch from half-gallon bottles to 1.75-liter bottles has provided an opportunity for opponents of conversion to charge to the general public that metrication will be used to cover unexplained price rises. In addition there clearly will be initial direct costs of conversion which are likely to fall unevenly over economic sectors. Finally, there will always be a portion of society which will be against change, no matter how gradual, especially when it is intimated that foreign influence is motivating the change.

To counteract the projected negative reaction to conversion, it is important that awareness concerning the metric system and its costs and benefits be communicated to the public. Coordination of conversion by the congressionally-established Metric Board should contribute to the mitigation of adverse consequence. Reduction in argumentation concerning the nuances of meanings and spellings of certain terms among the adherents of metrication would reduce impediments and ease acceptance of the metric system by the general public. It may be, however, that only with the coming of the age of the generation of students who have received their basic educational training by using the metric system will the United States really be able to accept and implement a complete transition to the International System of Units. This long range goal is both desirable and attainable.

Notes

1. See, for example, U.S., Department of Commerce, *Measures for Progress, A History of the National Bureau of Standards*, by R.C. Cochrane, National Bureau of Standards Special Publication 275, 1966.

2. U.S., Department of Commerce, *A History of the Metric System Controversy in the United States*, National Bureau of Standards Special Publication 345-10, 1971.

3. U.S., Department of Commerce, *The International Bureau of Weights and Measures 1875-1975*, edited by C.H. Page and P. Vigoureux, National Bureau of Standards Special Publication 420, 1975.

4. Minutes of the sixth meeting of the Consultative Committee on Units, May 17-18, 1978, Sèvres, France (in French).

5. U.S., Department of Commerce, *The International System of Units (SI)*, National Bureau of Standards Special Publication 330, 1977. See also appendix.

6. See, for example, "ASTM/IEEE Standard Metric Practice," ASTM E380-76/IEEE Std 268-1976, and "Legal Units of Measurement," Draft International Document OIML-1978, available from the International Organization for Legal Metrology, Rue Turgot, 75009 Paris, France.

7. "SI Units and Recommendations for the Use of their Multiples and of Certain Other Units," ISO 1000 (1973). International Organization for Standardization, Geneva, Switzerland.

8. H. Arthur Klein, *The World of Measurement* (New York: Simon & Schuster, 1974).

9. K. Boulding, "Numbers and Measurement on a Human Scale," chapter 3 of this volume.

10. Minutes of the sixth meeting of the Consultative Committee on Units.

11. D.F. Bartlett, "Natural Units: An Alternative to SI," chapter 4 of this volume.

12. S.E. Mallen, "Metric Fastener Overview," chapter 2 of this volume.

13. H.J. Milton, "Our Metric Future," *The Construction Specifier* (February 1979), p. 36.

14. Metric Conversion Act of 1975, Public Law 94-168.

15. Talk given by David Rubenstein, Deputy Assistant to the President for Domestic Affairs to the Americas. National Metric Council, April 3, 1979.

16. U.S., General Accounting Office, *Getting a Better Understanding of the Metric System — Implications if Adopted by the U.S.*, Report to the Congress by the Comptroller General, October 1978.

7.
The Three Dimensional Character of the Imperial System

(Editor's note: Newspaper accounts of the Houston meeting have encouraged some correspondence. The following letter by an English architect presents a new view of an old system.)

Dear Professor Bartlett,

You ask what I meant by the three-dimensional character of the imperial system.

I must confess to having little or nothing to offer in reply that has not been observed many times by better minds, John Quincy Adams for one, whom you quote so aptly.

Perhaps I may put it in personal terms. For me at least the imperial system is composed of recognizably solid objects. I call them objects because they present themselves in the mind, not as abstractions or ideas, but as comprehensible things. This is true whether one is speaking of capacity, weight, volume, mass or length. It follows that the basic units both increase through stages, and decrease by fractions, which are also recognizable as objects. The furlong, the quarter acre, the hundredweight are not abstractions.

I do not think I am being merely sentimental in this, out of affection for familiar, not to say time-hallowed names. I would be prepared, for the sake of general acceptability, to lose these if necessary: provided that the sense of interrelated solids was retained.

The imperial units, since they have evolved as practical tools of trade, and have been needed to work in harness one with another, as pounds per square inch, bushels per acre and so forth, have a recognizable family likeness and compatibility, as if all were descended from common progenitors.

To interpose a reminiscence: my father, who was a lawyer, could add up, or cast as they used to say, columns of £.s.d. in his head. To him they were not figures, they were interrelated pieces. (Oddly enough our people now regret the loss of £.s.d. They think, with some reason, that they have been defrauded by decimals.)

The metric system, by contrast, seems to me to be purely linear, an endless succession of marks on a scale on which no number has any greater significance than any other. It is true that one receives a command to halt at every tenth mark. But in a system in which no number appears to have any different character or essential being to any other it is hard to see any special reason for halting at ten. One stops dutifully, but only to proceed straight on down the track. Ten as a number has none of the solidity of twelve, nor its potential. Ten is like a wayside halt on a through line. Twelve is a major junction from which one has a choice of movement in many directions.

The key, one might say the parental, numbers seem to me to be three and four. Is one merely playing some kind of numbers game if one sees these and others as having identities, and indeed genders? How is one to explain one's sense of the significance of numbers, which seems to be as old as the earliest human stirrings amongst the flints and swamps, and to have been religious before it became practical? Why, one wonders, were there The Twelve? Did the number not mysteriously have to be twelve plus one, the one for luck, the famous baker's dozen, symbol both of generosity and of premonition?

Napoleon, needless to say, cared for none of these things. Nor certainly did the bureaucrats who formulated the SI system. And what, as you say, does it all matter in connection with the price of wheat in Chicago?

You are right to refer to packaging, which affords interesting examples of three dimensional number patterns, rather like the game Plato [a three-dimensional version of tic-tac-toe]. We have,

for example, a system of packing eggs which is so neat that I suspect it must be American. A tray of eggs, made of moulded card-
board, holds five times six eggs, equals two and a half dozen. Eight
of these trays can be stacked (any more and some eggs will break),
making twenty dozen. An egg box contains three stacks, or sixty
dozen in all. The metric system will just not function in this way.
It could easily have been some character in Chicago who thought
that one out.

Interestingly enough our own construction industry, which
has been semi-metricized for some time, is not finding the system
wholly helpful or accurate. (Many contractors transpose dimensions back to imperial to help site work.)

A dimension such as 11⅞ is unequivocal, neither more nor
less. It is solid. By contrast 11.875 is at best a mnemonic. It is an
abstraction. Depending on where along the scale it comes, it may
be rounded up to 11.88 or to 11.90 or (if in millimeters) to 12.00.
This rounding up and down to preferred dimensions is part of the
SI act. However, construction sites are a cat's cradle maze of
dimensions. As soon as one starts to pretend that a dimension
would look neater on a drawing if it were something else, and
when one then multiplies the discrepancy by a few structural bays
or whatever, and when one then allows for a degree of site bodg-
ing (to make the blessed thing work somehow), then one soon
gets into a Laurel and Hardy situation. Indeed I am sorry they are
not still with us: they could have had a lot of fun out of metrics.

You could of course argue that this is all just temporary teeth-
ing trouble. True enough maybe. But the fact that it happens is
enough to indicate the theoretical, committee-room nature of the
SI system.

I will say no more, since you have dealt with it, on the matter
of the fatuously enormous numbers required by the more complex calculations, nor on the terrifying errors awaiting the mis-
placement of a single decimal point. What we will lose, in adopting a system of abstractions, is of course the safety harness of the
ordinary mental arithmetic check, so readily to hand in imperial.

In conclusion, yes please do, you scientists, produce a new and
viable concept for one. May I assume that you could do this?

What then would be the next step? Presumably proponents of both metric (through all manner of by now entrenched interests) and imperial would then urge that the one system or the other be calibrated (or whatever would be the correct word) to the new unit or units. It is difficult to imagine how this intermediate, evolutionary stage would develop (one calls to mind the struggle to save Brunel's 7'0" Great Western Railway gauge). There would be some outcome eventually, even if it was the usual accommodating compromise. Then metric, if it won outright, would retain all its characteristics, simply transposed up or down, and all its serious and fundamental deficiencies. Imperial on the other hand could be adapted quite readily to a binary system, and (or am I wrong?) it could retain its tested structural fabric, what I have called its three-dimensional character.

In my innocence I can imagine that, having got this far, the problem might not prove to be all that impossibly hard from there on. Nomenclature would certainly not be the main consideration: but relationships would.

With kind regards,
Yours sincerely,

P.F. Bedford

Appendix :
The International
System of Units (SI)

I. INTRODUCTION

I.1 Historical note

In 1948 the 9th CGPM , by its Resolution 6, instructed the CIPM : "to study the establishment of a complete set of rules for units of measurement"; "to find out for this purpose, by official inquiry, the opinion prevailing in scientific, technical, and educational circles in all countries" and "to make recommendations on the establishment of a *practical system of units of measurement* suitable for adoption by all signatories to the Meter Convention."

The same General Conference also laid down, by its Resolution 7, general principles for unit symbols (see II.1.2, page 6) and also gave a list of units with special names.

The 10th CGPM (1954), by its Resolution 6, and the 14th CGPM (1971) by its Resolution 3, adopted as base units of this "practical system of units", the units of the following seven quantities: length, mass, time, electric current, thermodynamic temperature, amount of substance, and luminous intensity (see II.1, page 3).

The 11th CGPM (1960), by its Resolution 12, adopted the name *International System of Units*, with the international abbreviation SI, for this practical system of units of measurement and laid down rules for the prefixes (see III.1, page 10), the derived and supplementary units (see II.2, page 6 and II.3, page 9) and other matters, thus establishing a comprehensive specification for units of measurement.

In the present document the expressions "SI units", "SI prefixes", "supplementary units" are used in accordance with Recommendation 1 (1969) of the CIPM.

I.2 The three classes of SI units

SI units are divided into three classes:

> base units,
> derived units,
> supplementary units.

From the scientific point of view division of SI units into these three classes is to a certain extent arbitrary, because it is not essential to the physics of the subject.

Nevertheless the General Conference, considering the advantages of a single, practical, worldwide system for international relations, for teaching and for scientific work, decided to base the International System on a choice of seven well-defined units which by convention are regarded as dimensionally independent: the meter, the kilogram, the second, the ampere, the kelvin, the mole, and the candela (see II.1, page 3). These SI units are called *base units*.

The second class of SI units contains *derived units,* i.e., units that can be formed by combining base units according to the algebraic relations linking the corresponding quantities. Several of these algebraic expressions in terms of base units can be replaced by special names and symbols which can themselves be used to form other derivd units (see II.2, page 6).

Although it might be thought that SI units can only be base units or derived units, the 11th CGPM (1960) admitted a third class of SI units, called *supplementary units,* for which it declined to state whether they were base units or derived units (see II.3, page 9).

The SI units of these three classes form a coherent set in the sense normally attributed to the expression "coherent system of units".

The decimal multiples and sub-multiples of SI units formed by means of SI prefixes must be given their full name *multiples and sub-multiples of SI units* when it is desired to make a distinction between them and the coherent set of SI units.

II. SI UNITS

II.1 SI base units

1. Definitions

a.) The 11th CGPM (1960) replaced the definition of the meter based on the international prototype of platinum-iridium, in force since 1889 and amplified in 1927, by the following definition:

The meter is the length equal to 1 650 763.73 *wavelengths in vacuum of the radiation corresponding to the transition between the levels* $2p_{10}$ *and* $5d_5$ *of the krypton-86 atom.* (11th CGPM (1960), Resolution 6).

The old international prototype of the meter which was legalized by the 1st CGPM in 1889 is still kept at the International Bureau of Weights and Measures under the conditions specified in 1889.

b.) The 1st CGPM (1889) legalized the international prototype of the kilogram and declared: *this prototype shall henceforth be considered to be the unit of mass.*

The 3d CGPM (1901), in a declaration intended to end the ambiguity which existed as to the meaning of the word "weight" in popular usage, confirmed that the *kilogram is the unit of mass; it is equal to the mass of the international prototype of the kilogram* (see the complete declaration, p. 16).

This international prototype made of platinum-iridium is kept at the BIPM under conditions specified by the 1st CGPM in 1889.

c.) Originally the unit of time, the second, was defined as the fraction 1/86 400 of the mean solar day. The exact definition of "mean solar day" was left to astronomers, but their measurements have shown that on account of irregularities in the rotation of Earth, the mean solar day does not guarantee the desired accuracy. In order to define the unit of time more precisely the 11th CGPM (1960) adopted a definition given by the International Astronomical Union which was based on the tropical year. Experimental work had however already shown that an atomic standard of time-interval, based on a transition between two energy levels of an atom or a molecule, could be realized and reproduced much more accurately. Considering that a very precise definition of the unit of time of the International System, the second, is indispensable for the needs of advanced metrology, the 13th CGPM (1967) decided to replace the definition of the second by the following:

The second is the duration of 9 192 631 770 *periods of the radiation corresponding to the transition between the two hyperfine levels of the ground state of the cesium-133 atom.* (13th CGPM (1967), Resolution 1).

d.) Electric units, called "international", for current and resistance, had been introduced by the International Electrical Congress held in Chicago in 1893, and the definitions of the "international" ampere and the "international" ohm were confirmed by the International Conference of London in 1908.

Although it was already obvious on the occasion of the 8th CGPM (1933) that there was a unanimous desire to replace those "international" units by so-called "absolute" units, the official decision to abolish them was only taken by the 9th CGPM (1948), which adopted for the unit of electric current, the ampere, the following definition:

The ampere is that constant current which, if maintained in two straight parallel conductors of infinite length, of negligible circular cross section, and placed 1 meter apart in vacuum, would produce between these conductors a force equal to 2 × 10⁻⁷ newton per meter of length. (CIPM (1946), Resolution 2 approved by the 9th CGPM, 1948).

The expression "MKS unit of force" which occurs in the original text has been replaced here by "newton" adopted by the 9th CGPM (1948, Resolution 7).

e.) The definition of the unit of thermodynamic temperature was given in substance by the 10th CGPM (1954, Resolution 3) which selected the triple point of water as the fundamental fixed point and assigned to it the temperature 273.16 K by definition. The 13th CGPM (1967, Resolution 3) adopted the name *kelvin* (symbol K) instead of "degree Kelvin" (symbol °K) and in its Resolution 4 defined the unit of thermodynamic temperature as follows:

The kelvin, unit of thermodynamic temperature, is the fraction 1/273.16 of the thermodynamic temperature of the triple point of water. (13th CGPM (1967), Resolution 4).

The 13th CGPM (1967, Resolution 3) also decided that the unit kelvin and its symbol K should be used to express an interval or a difference of temperature.

Note.—In addition to the thermodynamic temperature (symbol T), expressed in kelvins, use is also made of Celsius temperature (symbol t) defined by the equation

$$t = T - T_0$$

where T_0 = 273.15 K by definition. The unit "degree Celsius" is equal to the unit "kelvin," but "degree Celsius" is a special name in place of "kelvin" for expressing Celsius temperature. A temperature interval or a Celsius temperature difference can be expressed in degrees Celsius as well as in kelvins.

f.) Since the discovery of the fundamental laws of chemistry, units of amount of substance called, for instance, "gram-atom" and "gram-molecule", have been used to specify amounts of chemical elements or compounds. These units had a direct connection with "atomic weights" and "molecular weights", which were in fact relative masses. "Atomic weights" were originally referred to the atomic weight of oxygen (by general agreement taken as 16). But whereas physicists separated isotopes in the mass spectrograph and attributed the value 16 to one of the isotopes of oxygen, chemists attributed that same value to the (slightly variable) mixture of isotopes 16, 17, 18, which was for them the naturally occurring element oxygen. Finally an agreement between the International Union of Pure and Applied Physics (IUPAP) and the International Union of Pure and Applied Chemistry (IUPAC) brought this duality to an end in 1959/60. Physicists and chemists have ever since agreed to assign the value 12 to the isotope 12 of carbon. The unified scale thus obtained gives values of "relative atomic mass".

It remained to define the unit of amount of substance by fixing the corresponding mass of carbon 12; by international agreement, this mass has been fixed at 0.012 kg, and the unit of the quantity, "amount of substance",[2] has been given the name *mole* (symbol mol).

Following proposals of IUPAP, IUPAC, and ISO, the CIPM gave in 1967, and confirmed in 1969, the following definition of the mole, adopted by the 14th CGPM (1971, Resolution 3) :

The mole is the amount of substance of a system which contains as many elementary entities as there are atoms in 0.012 kilogram of carbon 12.

Note. When the mole is used, the elementary entities must be specified and may be atoms, molecules, ions, electrons, other particles, or specified groups of such particles.

Note that this definition specifies at the same time the nature of the quantity whose unit is the mole.[2]

g.) The units of luminous intensity based on flame or incandescent filament standards in use in various countries were replaced in 1948 by the "new candle". This decision had been prepared by the International Commission on Illumination (CIE) and by the CIPM before 1937, and was promulgated by the CIPM at its meeting in 1946 in virtue of powers conferred on it in 1933 by the 8th CGPM. The 9th CGPM (1948) ratified the decision of the CIPM and gave a new international name, *candela* (symbol cd), to the unit of luminous intensity. The text of the definition of the candela, as amended in 1967, is as follows.

The candela is the luminous intensity, in the perpendicular direction, of a surface of 1/600 000 square meter of a blackbody at the temperature of freezing platinum under a pressure of 101 325 newtons per square meter. (13th CGPM (1967). Resolution 5).

---

[2] The name of this quantity, adopted by IUPAP, IUPAC, and ISO is in French "quantité de matière" and in English "amount of substance"; (the German and Russian translations are "Stoffmenge" and "количество вещества"). The French name recalls "quantitas materiae" by which in the past the quantity now called mass used to be known; we must forget this old meaning, for mass and amount of substance are entirely different quantities.

2. Symbols

The base units of the International System are collected in table 1 with their names and their symbols (10th CGPM (1954), Resolution 6; 11th CGPM (1960), Resolution 12; 13th CGPM (1967), Resolution 3; 14th CGPM (1971), Resolution 3).

TABLE 1

SI base units

Quantity †	Name	Symbol
length	meter	m
mass	kilogram	kg
time	second	s
electric current	ampere	A
thermodynamic temperature	kelvin	K
amount of substance	mole	mol
luminous intensity	candela	cd

† Translators' note:
"Quantity" is the technical word for measurable attributes of phenomena or matter.

The general principle governing the writing of unit symbols had already been adopted by the 9th CGPM (1948), Resolution 7, according to which:

Roman (upright) *type, in general lower case, is used for symbols of units; if however the symbols are derived from proper names, capital roman type is used* [for the first letter]. *These symbols are not followed by a full stop* (period).

Unit symbols do not change in the plural.

II.2 SI derived units

1. Expressions

Derived units are expressed algebraically in terms of base units by means of the mathematical symbols of multiplication and division. Several derived units have been given special names and symbols which may themselves be used to express other derived units in a simpler way than in terms of the base units.

Derived units may therefore be classified under three headings. Some of them are given in tables 2, 3, and 4.

TABLE 2

Examples of SI derived units expressed in terms of base units

Quantity	SI unit	
	Name	Symbol
area	square meter	m^2
volume	cubic meter	m^3
speed, velocity	meter per second	m/s
acceleration	meter per second squared	m/s^2
wave number	1 per meter	m^{-1}
density, mass density	kilogram per cubic meter	kg/m^3
current density	ampere per square meter	A/m^2
magnetic field strength	ampere per meter	A/m
concentration (of amount of substance)	mole per cubic meter	mol/m^3
specific volume	cubic meter per kilogram	m^3/kg
luminance	candela per square meter	cd/m^2

TABLE 3

SI derived units with special names

Quantity	SI unit Name	Symbol	Expression in terms of other units	Expression in terms of SI base units
frequency	hertz	Hz		s^{-1}
force	newton	N		$m \cdot kg \cdot s^{-2}$
pressure, stress	pascal	Pa	N/m^2	$m^{-1} \cdot kg \cdot s^{-2}$
energy, work, quantity of heat	joule	J	$N \cdot m$	$m^2 \cdot kg \cdot s^{-2}$
power, radiant flux	watt	W	J/s	$m^2 \cdot kg \cdot s^{-3}$
quantity of electricity, electric charge	coulomb	C	$A \cdot s$	$s \cdot A$
electric potential, potential difference, electromotive force	volt	V	W/A	$m^2 \cdot kg \cdot s^{-3} \cdot A^{-1}$
capacitance	farad	F	C/V	$m^{-2} \cdot kg^{-1} \cdot s^4 \cdot A^2$
electric resistance	ohm	Ω	V/A	$m^2 \cdot kg \cdot s^{-3} \cdot A^{-2}$
conductance	siemens	S	A/V	$m^{-2} \cdot kg^{-1} \cdot s^3 \cdot A^2$
magnetic flux	weber	Wb	$V \cdot s$	$m^2 \cdot kg \cdot s^{-2} \cdot A^{-1}$
magnetic flux density	tesla	T	Wb/m^2	$kg \cdot s^{-2} \cdot A^{-1}$
inductance	henry	H	Wb/A	$m^2 \cdot kg \cdot s^{-2} \cdot A^{-2}$
Celsius temperature (a)	degree Celsius	°C		K
luminous flux	lumen	lm		$cd \cdot sr^{(b)}$
illuminance	lux	lx	lm/m^2	$m^{-2} \cdot cd \cdot sr^{(b)}$
activity (of a radionuclide)†	becquerel	Bq		s^{-1}
absorbed dose, specific energy imparted, kerma, absorbed dose index	gray	Gy	J/kg	$m^2 \cdot s^{-2}$

(a) See page 4.
(b) In this expression the steradian (sr) is treated as a base unit.

† Translators' note: this term is more appropriate than the direct translation 'ionizing radiations' of the present French text.

TABLE 4

Examples of SI derived units expressed by means of special names

Quantity	SI unit Name	Symbol	Expression in terms of SI base units
dynamic viscosity	pascal second	Pa·s	$m^{-1} \cdot kg \cdot s^{-1}$
moment of force	newton meter	N·m	$m^2 \cdot kg \cdot s^{-2}$
surface tension	newton per meter	N/m	$kg \cdot s^{-2}$
power density, heat flux density, irradiance	watt per square meter	W/m²	$kg \cdot s^{-3}$
heat capacity, entropy	joule per kelvin	J/K	$m^2 \cdot kg \cdot s^{-2} \cdot K^{-1}$
specific heat capacity, specific entropy	joule per kilogram kelvin	J/(kg·K)	$m^2 \cdot s^{-2} \cdot K^{-1}$
specific energy	joule per kilogram	J/kg	$m^2 \cdot s^{-2}$
thermal conductivity	watt per meter kelvin	W/(m·K)	$m \cdot kg \cdot s^{-3} \cdot K^{-1}$
energy density	joule per cubic meter	J/m³	$m^{-1} \cdot kg \cdot s^{-2}$
electric field strength	volt per meter	V/m	$m \cdot kg \cdot s^{-3} \cdot A^{-1}$
electric charge density	coulomb per cubic meter	C/m³	$m^{-3} \cdot s \cdot A$
electric flux density	coulomb per square meter	C/m²	$m^{-2} \cdot s \cdot A$
permittivity	farad per meter	F/m	$m^{-3} \cdot kg^{-1} \cdot s^4 \cdot A^2$
permeability	henry per meter	H/m	$m \cdot kg \cdot s^{-2} \cdot A^{-2}$
molar energy	joule per mole	J/mol	$m^2 \cdot kg \cdot s^{-2} \cdot mol^{-1}$
molar entropy, molar heat capacity	joule per mole kelvin	J/(mol·K)	$m^2 \cdot kg \cdot s^{-2} \cdot K^{-1} \cdot mol^{-1}$
exposure (x and γ rays)	coulomb per kilogram	C/kg	$kg^{-1} \cdot s \cdot A$
absorbed dose rate	gray per second	Gy/s	$m^2 \cdot s^{-3}$

Note—The values of certain so-called dimensionless quantities, as for example refractive index, relative permeability or relative permittivity, are expressed by pure numbers. In this case the corresponding SI unit is the ratio of the same two SI units and may be expressed by the number 1.

Although a derived unit can be expressed in several equivalent ways by using names of base units and special names of derived units, the CIPM sees no objection to the use of certain combinations or of certain special names in order to distinguish more easily between quantities of the same dimension. For example, the hertz is used, instead of the reciprocal second, for frequency; and the newton meter, instead of the joule, for the moment of a force.

In the field of ionizing radiation, the becquerel is similarly used, instead of the reciprocal second, for activity; and the gray, instead of the joule per kilogram, for specific energy imparted, kerma, absorbed dose, and absorbed dose index.

2. Recommendations

The International Organization for Standardization (ISO) has issued additional recommendations with the aim of securing uniformity in the use of units, in particular those of the International System (see the series of International Standards ISO 31 and International Standard ISO 1000 of Technical Committee ISO/TC 12 "Quantities, units, symbols, conversion factors and conversion tables").

According to these recommendations:

a) The product of two or more units may be indicated in any of the following ways,

for example: $N \cdot m$, N.m, or N m.

b) A solidus (oblique stroke, /), a horizontal line, or negative powers may be used to express a derived unit formed from two others by division

for example: m/s, $\dfrac{m}{s}$ or $m \cdot s^{-1}$

c) The solidus must not be repeated on the same line unless ambiguity is avoided by parentheses. In complicated cases negative powers or parentheses should be used

for example: m/s^2 or $m \cdot s^{-2}$ *but not:* $m/s/s$
$m \cdot kg/(s^3 \cdot A)$ or $m \cdot kg \cdot s^{-3} \cdot A^{-1}$ $m \cdot kg/s^3/A$

II.3 SI supplementary units
The General Conference has not yet classified certain units of the International System under either base units or derived units. These SI units are assigned to the third class called "supplementary units", and may be regarded either as base units or as derived units.

For the time being this class contains only two, purely geometrical, units: the SI unit of plane angle, the *radian*, and the SI unit of solid angle, the *steradian* (11th CGPM (1960), Resolution 12).

TABLE 5

SI supplementary units

Quantity	SI unit	
	Name	Symbol
plane angle	radian	rad
solid angle	steradian	sr

The radian is the plane angle between two radii of a circle which cut off on the circumference an arc equal in length to the radius.
The steradian is the solid angle which, having its vertex in the center of a sphere, cuts off an area of the surface of the sphere equal to that of a square with sides of length equal to the radius of the sphere.

(International Standard ISO 31/I).

Supplementary units may be used to form derived units. Examples are given in table 6.

TABLE 6

Examples of SI derived units formed by using supplementary units

Quantity	SI units	
	Name	Symbol
angular velocity	radian per second	rad/s
angular acceleration	radian per second squared	rad/s^2
radiant intensity	watt per steradian	W/sr
radiance	watt per square meter steradian	$W \cdot m^{-2} \cdot sr^{-1}$

III. DECIMAL MULTIPLES AND SUB-MULTIPLES OF SI UNITS

III.1 SI prefixes

The 11th CGPM (1960, Resolution 12) adopted a first series of names and symbols of prefixes to form decimal multiples and submultiples of SI units. Prefixes for 10^{-15} and 10^{-18} were added by the 12th CGPM (1964, Resolution 8) and those for 10^{15} and 10^{18} by the 15th CGPM (1975, Resolution 10).

TABLE 7

SI prefixes

Factor	Prefix	Symbol	Factor	Prefix	Symbol
10^{18}	exa	E	10^{-1}	deci	d
10^{15}	peta	P	10^{-2}	centi	c
10^{12}	tera	T	10^{-3}	milli	m
10^{9}	giga	G	10^{-6}	micro	μ
10^{6}	mega	M	10^{-9}	nano	n
10^{3}	kilo	k	10^{-12}	pico	p
10^{2}	hecto	h	10^{-15}	femto	f
10^{1}	deka	da	10^{-18}	atto	a

III.2 Recommendations

ISO recommends the following rules for the use of SI prefixes:

a) Prefix symbols are printed in roman (upright) type without spacing between the prefix symbol and the unit symbol.

b) An exponent attached to a symbol containing a prefix indicates that the multiple or sub-multiple of the unit is raised to the power expressed by the exponent,

for example: $1\text{ cm}^3 = (10^{-2}\text{ m})^3 = 10^{-6}\text{ m}^3$

$1\text{ cm}^{-1} = (10^{-2}\text{ m})^{-1} = 10^{2}\text{ m}^{-1}$

$1\ \mu\text{s}^{-1} = (10^{-6}\text{ s})^{-1} = 10^{6}\text{ s}^{-1}$

c) Compound prefixes, formed by the juxtaposition of two or more SI prefixes, are not to be used.

for example: 1 nm   *but not:* 1 mμm

III.3 The kilogram

Among the base units of the International System, the unit of mass is the only one whose name, for historical reasons, contains a prefix. Names of decimal multiples and sub-multiples of the unit of mass are formed by attaching prefixes to the word "gram" (CIPM (1967), Recommendation 2).

IV. UNITS OUTSIDE THE INTERNATIONAL SYSTEM

IV.1 Units used with the International System

The CIPM (1969) recognized that users of SI will wish to employ with it certain units not part of it, but which are important and are widely used. These units are given in table 8. The combination of units of this table with SI units to form compound units should be restricted to special uses in order to not lose the advantage of the coherence of SI units.

TABLE 8

Units in use with the International System

Name	Symbol	Value in SI unit
minute	min	1 min $= 60$ s
hour[a]	h	1 h $\; = 60$ min $= 3\,600$ s
day	d	1 d $\; = 24$ h $= 86\,400$ s
degree	°	$1° \; = (\pi/180)$ rad
minute	′	$1' \; = (1/60)° = (\pi/10\,800)$ rad
second	″	$1'' = (1/60)' = (\pi/648\,000)$ rad
liter[b]	l	$1 \; l \; = 1$ dm$^3 = 10^{-3}$ m^3
metric ton	t	1 t $= 10^3$ kg

[a] The symbol of this unit is included in Resolution 7 of the 9th CGPM (1948).
[b] This unit and its symbol were adopted by CIPM in 1879 (Procès-Verbaux CIPM, 1879, p. 41). For the symbol for liter, when there is a risk of confusion between the letter l and the number 1, one may use the abbreviation "ltr" or write "liter" in full (CIPM, 1976). The present definition of the liter is in Resolution 6 of the 12th CGPM (1964).

It is likewise necessary to recognize, outside the International System, some other units which are useful in specialized fields, because their values expressed in SI units must be obtained by experiment, and are therefore not known exactly (table 9).

TABLE 9

Units used with the International System whose values in SI units are obtained experimentally

Name	Symbol	Definition
electronvolt	eV	[a]
unified atomic mass unit	u	[b]
astronomical unit	[c]	[c]
parsec	pc	[d]

[a] 1 electronvolt is the kinetic energy acquired by on electron in passing through a potential difference of 1 volt in vacuum; 1 eV $= 1.602\,19 \times 10^{-19}$ J approximately.
[b] The unified atomic mass unit is equal to the fraction $1/12$ of the mass of an atom of the nuclide ^{12}C; 1 u$=1.660\,57 \times 10^{-27}$ kg approximately.
[c] This unit does not have an international symbol; abbreviations are used, for example, AU in English, UA in French, AE in German, a.e. Д in Russian, etc. The astronomical unit of distance is the length of the radius of the unperturbed circular orbit of a body of negligible mass moving round the Sun with a sidereal angular velocity of 0.017 202 098 950 radian per day of 86 400 ephemeris seconds. In the system of astronomical constants of the International Astronomical Union (1976), the value adopted is 1 AU $= 149\,597.870 \times 10^6$ m.
[d] 1 parsec is the distance at which 1 astronomical unit subtends an angle of 1 second of arc; we thus have approximately, 1 pc $= 206\,265$ AU $= 30\,857 \times 10^{12}$ m.

IV.2 Units accepted temporarily

In view of existing practice the CIPM (1969) considered it was preferable to keep temporarily, for use with those of the International System, the units listed in table 10.

TABLE 10

*Units to be used temporarily with the
International System*

Name	Symbol	Value in SI unit
nautical mile[a]		1 nautical mile = 1 852 m
knot		1 nautical mile per hour =
		(1852/3600) m/s
ångström	Å	1 Å = 0.1 nm = 10^{-10} m
are[b]	a	1 a = 1 dam^2 = 10^2 m^2
hectare[b]	ha	1 ha = 1 hm^2 = 10^4 m^2
barn[c]	b	1 b = 100 fm^2 = 10^{-28} m^2
bar[d]	bar	1 bar = 0.1 MPa = 10^5 Pa
standard atmosphere[e]	atm	1 atm = 101 325 Pa
gal[f]	Gal	1 Gal = 1 cm/s^2 = 10^{-2} m/s^2
curie[g]	Ci	1 Ci = 3.7 × 10^{10} Bq
röntgen[h]	R	1 R = 2.58 × 10^{-4} C/kg
rad[i]	rad	1 rad = 1 cGy = 10^{-2} Gy

[a] The nautical mile is a special unit employed for marine and aerial navigation to express distances. The conventional value given above was adopted by the First International Extraordinary Hydrographic Conference, Monaco, 1929, under the name "International nautical mile".

[b] This unit and its symbol were adopted by the CIPM in 1879 (*Procès-Verbaux CIPM*, 1879, p. 41). USA Editors' note: In recommended USA practice, hectare would appear in table 8.

[c] The barn is a special unit employed in nuclear physics to express effective cross sections.

[d] This unit and its symbol are included in Resolution 7 of the 9th CGPM (1948).

[e] Resolution 4 of 10th CGPM (1954).

[f] The gal is a special unit employed in geodesy and geophysics to express the acceleration due to gravity.

[g] The curie is a special unit employed in nuclear physics to express activity of radionuclides (12th CGPM (1964), Resolution 7).

[h] The röntgen is a special unit employed to express exposure of x or γ radiations.

[i] The rad is a special unit employed to express absorbed dose of ionizing radiations. When there is risk of confusion with the symbol for radian, rd may be used as the symbol for rad.

IV.3 CGS units

The CIPM considers that it is in general preferable not to use, with the units of the International System, CGS units which have special names.[3] Such units are listed in table 11.

TABLE 11

CGS units with special names

Name	Symbol	Value in SI unit
erg[a]	erg	$1 \text{ erg} = 10^{-7} \text{ J}$
dyne[a]	dyn	$1 \text{ dyn} = 10^{-5} \text{ N}$
poise[a]	P	$1 \text{ P} = 1 \text{ dyn·s/cm}^2 = 0.1 \text{ Pa·s}$
stokes	St	$1 \text{ St} = 1 \text{ cm}^2/\text{s} = 10^{-4} \text{ m}^2/\text{s}$
gauss[b]	Gs, G	$1 \text{ Gs corresponds to } 10^{-4} \text{ T}$
oersted[b]	Oe	$1 \text{ Oe corresponds to } \dfrac{1000}{4\pi} \text{ A/m}$
maxwell[b]	Mx	$1 \text{ Mx corresponds to } 10^{-8} \text{ Wb}$
stilb[a]	sb	$1 \text{ sb} = 1 \text{ cd/cm}^2 = 10^4 \text{ cd/m}^2$
phot	ph	$1 \text{ ph} = 10^4 \text{ lx}$

[a] This unit and its symbol were included in Resolution 7 of the 9th CGPM (1948).
[b] This unit is part of the so-called "electromagnetic" 3-dimensional CGS system and cannot strictly speaking be compared to the corresponding unit of the International System, which has four dimensions when only electric quantities are considered.

[3] The aim of the International System of Units and of the recommendations contained in this document is to secure a greater degree of uniformity, hence a better mutual understanding of the general use of units. Nevertheless in certain specialized fields of scientific research, in particular in theoretical physics, there may sometimes be very good reasons for using other systems or other units.
Whichever units are used, it is important that the *symbols* employed for them follow current international recommendations.

IV.4 Other units

As regards units outside the International System which do not come under sections IV.1, 2, and 3, the CIPM considers that it is in general preferable to avoid them, and to use instead units of the International System. Some of those units are listed in table 12.

TABLE 12

Other units generally deprecated

Name	Value in SI unit
fermi	1 fermi $= 1$ fm $= 10^{-15}$ m
metric carat[a]	1 metric carat $= 200$ mg $= 2 \times 10^{-4}$ kg
torr	1 torr $= \dfrac{101\ 325}{760}$ Pa
kilogram-force (kgf)	1 kgf $= 9.806\ 65$ N
calorie (cal)	1 cal $= 4.186\ 8$ J[b]
micron (μ) [c]	1 $\mu = 1\ \mu$m $= 10^{-6}$ m
x unit[d]	
stere (st) [e]	1 st $= 1$ m^3
gamma (γ)	1 $\gamma = 1$ nT $= 10^{-9}$ T
γ [f]	1 $\gamma = 1\ \mu$g $= 10^{-9}$ kg
λ [g]	1 $\lambda = 1\ \mu$l $= 10^{-6}$ l $= 10^{-9}$ m^3

[a] This name was adopted by the 4th CGPM (1907, pp. 89–91) for commercial dealings in diamonds, pearls, and precious stones.

[b] This value is that of the so-called "IT" calorie (5th International Conference on Properties of Steam, London, 1956).

[c] The name of this unit and its symbol, adopted by the CIPM in 1879 (*Procès-Verbaux CIPM*, 1879, p. 41) and retained in Resolution 7 of the 9th CGPM (1948) were abolished by the 13th CGPM (1967, Resolution 7).

[d] This special unit was employed to express wavelengths of x rays; 1 x unit $= 1.002 \times 10^{-4}$ nm approximately.

[e] This special unit employed to measure firewood was adopted by the CIPM in 1879 with the symbol "s" (*Procès-Verbaux CIPM*, 1879, p. 41). The 9th CGPM (1948, Resolution 7) changed the symbol to "st".

[f] This symbol is mentioned in *Procès-Verbaux CIPM*, 1880, p. 56.

[g] This symbol is mentioned in *Procès-Verbaux CIPM*, 1880, p. 30.